ANATOMY OF GENE REGULATION

A Three-Dimensional Structural Analysis

Determination of three-dimensional structures has revealed astonishing snapshots of molecules in action. No longer simple line drawings on a page, molecular structures can now be viewed in full-figured glory, often in color and even with interactive possibilities.

Anatomy of Gene Regulation is the first book to present the parts and processes of gene regulation at the three-dimensional level. Vivid structures of nucleic acids and their companion proteins are revealed in full-color, three-dimensional form. Beginning with a general introduction to three-dimensional structures, the book looks at the organization of the genome, the structure of DNA, DNA replication and transcription, splicing, protein synthesis, and ultimate protein death. The text discusses genetics and structural mechanics throughout. The concise and unique synthesis of information offers insight into gene regulation and into the development of methods to interfere with regulation at diseased states.

This textbook is appropriate for both undergraduate and graduate students in genetics, molecular biology, structural biology, and biochemistry courses.

Panagiotis A. Tsonis is Professor in the Department of Biology at the University of Dayton.

To my daughters
Isidora and Sol

ANATOMY OF GENE REGULATION

A Three-Dimensional Structural Analysis

PANAGIOTIS A. TSONIS
University of Dayton

CAMBRIDGE
UNIVERSITY PRESS

CAMBRIDGE
UNIVERSITY PRESS

Shaftesbury Road, Cambridge CB2 8EA, United Kingdom

One Liberty Plaza, 20th Floor, New York, NY 10006, USA

477 Williamstown Road, Port Melbourne, VIC 3207, Australia

314–321, 3rd Floor, Plot 3, Splendor Forum, Jasola District Centre, New Delhi – 110025, India

103 Penang Road, #05–06/07, Visioncrest Commercial, Singapore 238467

Cambridge University Press is part of Cambridge University Press & Assessment,
a department of the University of Cambridge.

We share the University's mission to contribute to society through the pursuit of
education, learning and research at the highest international levels of excellence.

www.cambridge.org
Information on this title: www.cambridge.org/9780521804745

© Cambridge University Press & Assessment 2003

First published 2003

A catalogue record for this publication is available from the British Library

Library of Congress Cataloging-in-Publication data
Tsonis, Panagiotis A.
Anatomy of gene regulation : a three-dimensional structural analysis / Panagiotis A. Tsonis.
    p.   cm.
Includes bibliographical references and index.
ISBN 0-521-80030-7 – ISBN 0-521-80474-4 (pb.)
1. Genetic regulation.   2. Nucleic acids – Structure.   3. Proteins – Structure.
4. Three-dimensional imaging in biology.   I. Title.
QH450.T78   2002
572.8′65 – dc21   2001037406

ISBN   978-0-521-80030-3   Hardback
ISBN   978-0-521-80474-5   Paperback

Contents

Preface

What is in the structure? It is, of course, quality!

The determination of the three-dimensional (3-D) structure of DNA in 1953 heralded the beginning of molecular biology. At the same time, we saw one of the first examples of how the 3-D structure of a biomolecule reveals its function. The 3-D structure of DNA immediately suggested how the genetic information is passed to the progeny. Eventually, the discovery of DNA structure led to the understanding of how genetic information accounts for the final product, which is protein synthesis. For the past 40 years, research in molecular biology has led to the identification of a cascade in gene regulation from its packing into chromosomes to transcription, splicing, modifications, protein synthesis, and, finally, the death of proteins. Eventually, knowledge of the mechanisms involved in these events led to manipulation of genes, recombinant DNA, and cloning technology, all of which helped us grasp the function of genes and their role in the study of differentiation, development, and diseases.

As the major players at all the different levels of gene regulation were discovered, it became apparent that the final mechanisms will be best revealed when we can observe the action of enzymes and genes at the 3-D level. Information gathered by biochemical and molecular experiments could identify the function of an enzyme, say the role of DNA polymerase in replication or the role of helicase in unwinding DNA. However, only by observing the process at the 3-D level can we visualize and fathom the mechanism by which the enzyme possesses such activity. In fact, we can visualize enzyme action not only at the molecular level but also at the atomic level.

All this was possible because of the development of technology to determine the 3-D structure of biological molecules. In this sense, molecular biology has given rise to atomic biology where interactions between biomolecules and mechanisms can be resolved at the atomic level. These techniques have become better over the years and undoubtedly will become even better. So far, the determination of 3-D structures has revealed astonishing snapshots of molecules in action and has made secrets known. We are able now to virtually study the anatomy of molecular events.

In this book, I present the story of gene regulation with emphasis at the 3-D level, thus the title "Anatomy of Gene Regulation." Like surgeons (molecular surgeons), we can now open the nucleus and other organelles and, with extraordinary glasses, see the mechanisms unfolding at the atomic level. I have elected to begin our discussion with the packing (organization) of DNA in the nucleus and then to explore DNA replication, transcription, and splicing, followed by RNA modification. Next, we will follow the path of mRNA to the cytoplasm and its decoding during protein synthesis. The final chapter will consider the birth and death of proteins, which is the end of the regulation process.

Why did I undertake such a task? Even though I am not a structural biologist, I have been teaching molecular biology for nearly 15 years. Throughout my teaching career, I realized that most of the molecular biology textbooks were incredibly massive, full of more information than can be covered in one semester. Furthermore, textbooks in different disciplines overlap considerably. Most textbooks on microbiology, genetics, molecular or cellular biology, and even developmental biology contain chapters on the basic aspects of gene regulation. Students, therefore, are exposed to the same contents thoughout their education. I believe that, at least in the study of molecular biology, some changes are warranted. Consequently, I slowly started incorporating the 3-D aspects. I soon realized that my course was becoming more unique and exciting to the students. Finally, I decided that a molecular biology course that emphasized 3-D aspects must be beneficial to both undergraduate and graduate students. Determining the 3-D structure of a mechanism is the ultimate level that we should go to if we are to understand the mechanism. Therefore, all disciplines dealing with cellular and molecular events will eventually need to study them at the 3-D level.

Producing this text was very challenging. For every structure presented in the book, I read the corresponding paper(s) very carefully and extracted only the information that would produce a sequence that flows well and is accurate. Therefore, my book is by design short and concise. It focuses on the 3-D aspects of gene regulation only. However, it does not leave out information pertinent to molecular biology when structures are not involved. I intend to focus on the 3-D aspects of gene regulation and not to overwhelm the reader with details that can be found in many other general textbooks and courses. Each chapter is preceded by a section, which I call primer. This section outlines the general plan of the sequence that I follow in the chapter. It should help keep the reader focused on the main aim of the text.

Even a book on 3-D structures can be overwhelming. For example, the structure of many different DNA polymerases must be solved. Likewise, we must know the 3-D structure of numerous transcriptional factors. Obviously, reporting all of them in a book would be tedious for the reader. Therefore, I have filtered the information and concentrated on a series of structures that best represent the desired style of the book and its aim, which is to acquaint the readers with the 3-D aspects of gene regulation. At the same time, I have tried not to compromise quality and accuracy. The structures are presented in different ways using different models. This technique is deliberate because some aspects of a 3-D structure are better depicted by one model than by others.

Such a synthesis has not been attempted before, so I am sure that there will be supporters and critics. I do count on both to provide me with valuable comments on this project so that can improve it in the future.

Finally, I am grateful to numerous colleagues who have furnished me with the figures that are included in the book. Without their help and co-operation, my book would have been impossible to produce. The reference list is not extensive and is concentrated mainly on the papers that deal with the 3-D structures. General references on standard molecular biology can be found in several excellent textbooks, which are cited in the reference section.

CREDITS

Several databases, which have cataloged available three-dimensional structures of proteins, were a big help. Throughout the book numerous images have been used from these databases. These databases follow:

Berman, H. M., Olson, W. K., Beveridge, D. L., Westbrook, J., Gelbin, A., Demesy, T., Hsieh, S-H., Srinivasan, A. R., and Schneider, B. (1992). The nucleic acid database: A comprehensive relational database of three-dimensional structures of nucleic acids (NDB). *Biophys. J.* 63: 751–9. http://ndbserver.rutgers.edu/NDB/NDBATLAS/

Berman, H. M., Westbrook, J., Zeng, Z., Gilliland, G., Bhat, T. N., Weissing, H., Shindyalov, I. N., and Bourne, P. E. (2000). The Protein Data Bank (PDB). Nucleic Acids Res. 28: 235–42. www.rcsb.org/pdb

Reichert, J., Jabs, P., Slickers, J., and Suhnel, J. (2000). The IMB Jena Image Library of Biological Macromolecules. Nucleic Acids Res. 28: 246–9. www.imb-jena.de

CHAPTER ONE

A General Introduction to 3-D Structures

——— — – – –

—————————————

PRIMER The three-dimensional structure of nucleic acids and proteins as it pertains to the mechanisms involved in gene regulation is the major focus of this book. Therefore, the reader will encounter many 3-D structures. The first chapter of the book presents the very basic ideas behind the three-dimensional aspects of biomolecules. The first part deals with the techniques used to determine 3-D structures. The presentation is virtually for the layperson. Then the basic structural elements found in proteins are examined. Having done this, we examine a particular 3-D structure (that includes both DNA and protein) presented with different modeling. This exercise will help you to become familiar with the different ways that scientists present their 3-D structures. We use different models because one aspect of structure and function can be better represented with one model, whereas another aspect is more suited to a different model.

This book deals with the three-dimensional aspects of gene regulation. The reader will encounter numerous three-dimensional structures, but this should not scare anybody away. Unfamiliar readers might think that interpreting these structures is difficult, but this is not true. All we need is a basic introduction into the three-dimensional aspects of proteins and nucleic acids and the way that it can be represented. The basic 3-D structure of a protein can be reduced to two elements: the alpha helix and the beta strand (and loops that connect them). The complicated 3-D structure of a protein is a combination of several of these elements. Also, depending on the presentation, the alpha helix or the beta strand might be shown with different styles. To get started, let us review the main elements of the 3-D structures, the different representation styles, and the basic methods used in the determination of the three-dimensional structures.

Two methods are generally used to determine the 3-D structure of a biomolecule (nucleic acid or protein). One method is Nuclear Magnetic Resonance (NMR) spectroscopy, and the other is X-ray diffraction (or X-ray crystallography). NMR uses properties of the atomic nuclei to determine how closely they are positioned. The so-called nuclear Overhauser effect (NOE) is a nuclear relaxation effect. This intensity is a measure of the distance between two nuclei that are close together. The two nuclei might be far apart in the primary sequence, but they could be close in 3-D because the protein is folded. Gathering data from all atoms enables the researcher to create a 3-D model of the molecule under investigation. NOE(s) are detected by NOE spectroscopy (NOESY) NMR experiments. The intensity of NOESY determines the actual distance between two nuclei. A strong intensity indicates that the two nuclei are 3 Å apart, a medium intensity measures less than 4 Å, and weak intensity is less than 6 Å. Because structure determination by NMR is in solution, ends or loops of proteins, which are flexible, are sometimes not solved well. For this, more than 20 calculated structures should be received and superimposed. At this point, we should be able to see the regions that are not defined well. Finally, based on all calculated structures, an average structural model can be produced. These superimposed structures appear throughout the text. One limitation of this method is that it can resolve structures of small proteins (about 30 kDa). However, a few proteins of about 50 kDa have also been solved, and future developments might push these limits. Also, because NMR determines structures in solution, the protein should be stable in solution.

The other method, using X-ray diffraction, can be applied to large molecules or even complexes of them. When using X-ray diffraction, the protein must be crystallized. The crystal is then exposed to X-rays, and a picture is received on a film where the diffracted light from the crystal produces patterns, depending on the 3-D structure of the protein. For example, the celebrated 3-D structure of DNA, which is a periodic pattern, produced spots on the film, which were symmetrically arranged. This symmetry led Watson and Crick to deduce that the DNA must have two periodicities, one from base to base and the other every helical turn (nearly every 10 bases; see Chapter 2). Obviously, most complicated 3-D structures, such as the ones found in proteins, would produce a more elaborate pattern on the film, but algorithms and techniques have been developed to put these patterns into a 3-D structure. X-ray diffraction would provide very clear 3-D solutions and does not have the limitations with the flexible regions as in NMR. The only limitation is that not all proteins can be crystallized efficiently. When the same structure has been solved with both NMR and X-ray diffraction, the results usually match very well, indicating that both methods are quite reliable.

Let us now familiarize ourselves with the basic structures in a protein. As noted earlier, the primary amino acid sequence can assume either a helical or a beta strand conformation. Some amino acids are more likely than others to be in an alpha helix, and the same is true for amino acids found in beta strands. First, we will examine the basic structure of an amino acid and the peptide bond. All amino acids have a central carbon, C_α, to which a hydrogen atom, NH_2 (amino group), and COOH (carboxyl group) are attached. What discriminates the 20 different amino acids is the side chain, R, which is attached to the central carbon atom (Figure 1.1A). Amino acids are joined via the peptide bond to create polypeptides (Figure 1.1B). When amino acids are arranged in an alpha helix, there is hydrogen bonding between the C=O of a residue and

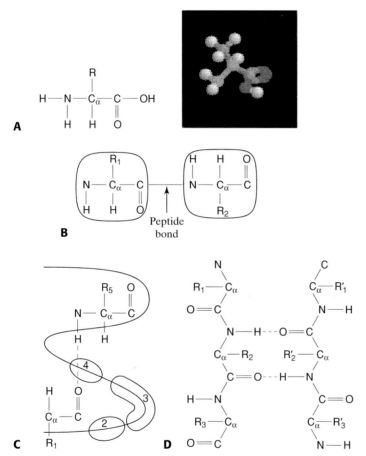

Figure 1.1. A: The basic chemical structure of an amino acid, indicating the standard H, NH_2, and COOH groups. R is the side chain that can vary in different amino acids. In the ball-and-stick representation, R is a CH_3 group and the amino acid is alanine. C is gray with the C_α cyan, N is blue, oxygen is red, and H is white. **B:** A dipeptide showing the creation of the peptide bond. **C:** An illustration of alpha helix. Note that residue 1 and residue 5 interact via hydrogen bonding (dashed line) using their C=O and NH groups, respectively. **D:** Two antiparallel beta strands creating a beta sheet via hydrogen bonding (dashed line) using an NH group from one strand and a C=O group from another. From F. R. Gorga, Protein Data Bank (PDB), Nucleic Acids Res. 28: 235–42.

the NH of another residue four positions away. In other words, there would be hydrogen bonds between residue 1 and 5, 2 and 6, and so on (Figure 1.1C). The alpha helix has 3.6 residues per turn, but variations exist with hydrogen bonds to residue $n + 5$ (pi helix) or $n + 3$ (3_{10} helix). Most of these helices are found at the end of alpha helices. The alpha helices are usually depicted as ribbons or cylinders in the 3-D structure of a protein.

The beta strand, and the resulting beta sheets from their interaction, is the second major element found in proteins. A beta strand contains 5 to 10 amino acids, which are in almost fully extended conformation. Interactions with adjacent beta strands can form a beta sheet. These interactions involve hydrogen bonding between the C=O of one strand and the NH group of another (Figure 1.1D). From such a configuration, we can see that the beta strands are pleated with C_α atoms successively above or below the plane of a sheet. The side chains follow this pattern as well. A beta sheet is called parallel when the strands run in the same direction or antiparallel when they do not. The example in Figure 1.1D is an antiparallel beta sheet. The beta strands are usually represented as arrows in the 3-D structure of a protein with the arrowhead pointing to the direction (N → C).

This book contains numerous structures that are represented as different models. This approach is deliberate because some models can show a particular feature much better than others. To illustrate, Figure 1.2 presents a particular structure using four different models. The structure shows the interaction between the paired domain of the activator pax-6 with DNA. It is a good example

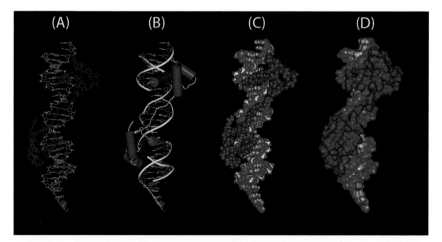

Figure 1.2. Different models of pax-6 bound to DNA. **A:** Ball-and-stick diagram. Phosphates are yellow, the sugar moiety is blue, nucleotide bases are gray, and pax-6 is red. **B:** Same as in A, but the helices of pax-6 are shown as red cylinders, and the connecting parts, as green strings. The DNA phosphates are yellow and have been traced to highlight the DNA. **C:** CPK (space-filling) model with same colors as in A. **D:** Surface representation with the same colors as in A. Images generated by E. Fuentes; Xu et al., Genes Develop. 13: 1263–75 (1999).

because we can examine both a nucleic acid and a protein. In Figure 1.2A, we can see the so-called ball-and-stick diagram, where atoms or groups are represented by balls (usually different colors; see also the model of alanine in Figure 1.1) and connected by sticks. In Figure 1.2B, we can see the same model as in Figure 1.2A, but the protein helices are represented as cylinders here. Helices can also be represented as ribbons. Both cylinder and ribbon models are used throughout this book. In Figure 1.2C, we can see the so-called CPK (Corey-Pauling-Kultun) model or spacefill (filling) model, which shows the surface of each atoms or group. Finally, in Figure 1.2D, we can see the solvent surface of the structure. This looks like the CPK model, but it represents the surface of the whole molecule instead of showing the surface of the atom or a group. This model is used mostly to represent the potential of a molecule, with red representing negative electrostatic potential and blue, positive. Variations of these models exist, but these models are the most common ones.

CHAPTER TWO

The Higher Organization of the Genome

——— ‒ ‒ ‒
_____

PRIMER DNA is not free in the nucleus; it is bound by the proteins that package it. This fact is important because DNA must be accommodated in a small place. Also, as we will see in Chapter 6, the packaging provides a level of transcriptional control. In this chapter, the goal is to become familiar with the proteins that are involved in packaging and their effects on DNA. The first part of the chapter examines the 3-D structure of the nucleosome and its components, the histones. This structure is the highest order of DNA packaging. Other proteins (nonhistone), however, have the ability to bend DNA and, therefore, provide another degree of organization. Such architectural proteins are important because linear DNA is not very efficient when it comes to regulation. Regulatory elements are often separated by long distances, and they must come close for interaction. Also, bending DNA makes it more accessible for interactions with proteins. These nonhistone proteins possess different structural motifs that mediate interaction with DNA. Their structural characteristics and effects on DNA are presented. The importance of these architectural proteins is also stressed in later chapters.

PACKING DNA IN THE CHROMOSOMES

The genetic material, DNA, is packaged into chromosomes as chromatin, which is the DNA and chromosomal proteins. The way that this packaging is achieved is suitable for DNA accommodation in the nucleus. It also most likely plays a role in DNA regulation. The packing of DNA is characterized by coils, loops, and coils within the loops. Such an organization guarantees that the DNA is packed in an orderly fashion, making use of less space in the nucleus and, at the same time, avoiding extensive knotting, which could inhibit the

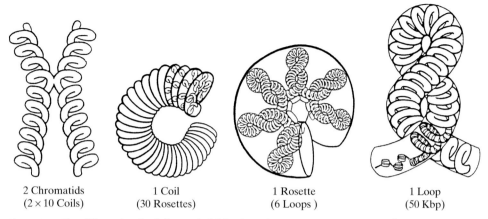

2 Chromatids	1 Coil	1 Rosette	1 Loop
(2 × 10 Coils)	(30 Rosettes)	(6 Loops)	(50 Kbp)

Figure 2.1. The different levels of chromatin folding into chromosomes. From P. R. Walker, EMBO J. 9: 1319–27 (1990).

necessary processes of replication and transcription. There are several levels of chromosomal folding. Starting from the higher level, we can observe that the chromosomes are coiled and that there appear to be nearly 10 coils per sister chromatid. Each coil is folded into rosettes, with 30 rosettes making up a coil. Each rosette is folded up into six loops with each loop containing up to 50 kbp of DNA. The loops are made up of 30-nm fiber, which is called solenoid. Each solenoid consists of 12 of the simplest units of packaging, the nucleosomes. At this level, we can observe that the DNA strands wrap around a proteinous core structure (Figure 2.1).

The proteins that make up a nucleosome are called histones, namely H2A, H2B, H3, and H4. The whole structure is an octamer of these histones and is constructed of two H2A-H2B and two H3-H4 dimers. The DNA (146 bp) wraps the histone complex in two turns (Figure 2.2).

The portion of the DNA that connects different nucleosomes depends on the species but is nearly 60 bp long and is bound by a different histone, H1. It seems that H1 binds DNA in such a fashion that a zigzag arrangement of the nucleosomes is possible. This arrangement is necessary for the nucleosomes to form the solenoids and has been observed in electron microscopy (Figure 2.3).

Figure 2.2. The nucleosome consisting of a histone complex (tube) and the wrapping DNA.

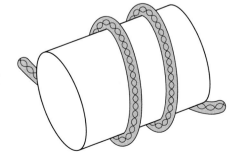

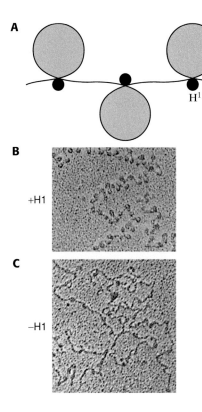

A

H¹

B

+H1

C

−H1

Figure 2.3. **A:** H1 binds free DNA. **B** and **C:** Binding of chromatin as observed by electron microscopy with (**B**) or without (**C**) H1. Note that the zigzag arrangement disappears in C. B and C reproduced from A. Klug, J. Cell Biol. 83: 410 (Nov. 1979), with copyright permission of the Rockefeller University Press.

THE THREE-DIMENSIONAL STRUCTURE OF THE NUCLEOSOME

The higher order of folding has not been observed through 3-D images, and speculation of the packaging order has been based on electron microscopy and biochemical studies. Therefore, our view of the higher order might change in the future when more accurate 3-D images will be available. However, the 3-D structure of the nucleosome has been solved at a level where accurate atomic interactions can be observed. These images have provided unique insights on the organization and structure of nucleosomes. As mentioned, the nucleosome is composed of a histone octamer wrapped by 146 bp of DNA. The specific arrangement of histones is characterized by the two H2A-H2B and two H3-H4 dimers. Before we examine the structure of the nucleosome, however, let us become familiar with the basic 3-D structure of histones. Each histone is composed of three helices that form the characteristic histone fold (Figure 2.4), and two tails with no particular structure.

Interactions between DNA and proteins can be seen in Figure 2.5, which is a cross section of a nucleosome (one H2A-H2B and one H3-H4 dimer with one turn of DNA). Several important features can be observed in Figure 2.5. H3 specifically binds DNA at the entrance and exit site. The wrapping DNA is contacted every 10 bases, at intervals at minor grooves that face the proteins. The phosphate backbone of the DNA is contacted by main-chain atoms of

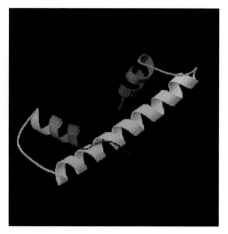

Figure 2.4. The characteristic structure of histones indicating the histone fold. The histone fold is characterized by one long alpha helix flanked on each side by a loop and a shorter alpha helix. This fold is for *Methanothermus ferridus* H3. From U. Heinemann, PDB file 1 HTA, K. Decanniere et al., J. Mol. Biol. 303: 35–47 (2000).

the proteins. This interaction is not sequence specific (because all different sequences of DNA wrap the nucleosomes) and is necessary to keep DNA tightly bound. Arginine residues penetrate all 14 minor grooves (7 in the cross section) and face the core of the protein arrangement.

The helical periodicity (average number of bases per helical turn of DNA) is 10.6. However, around the nucleosome, the periodicity is 10.2. In addition to

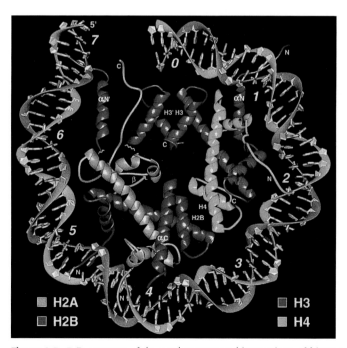

Figure 2.5. 3-D structure of the nucleosome and interactions of histones and DNA. From T. J. Richmond, Nature 389: 251–60 (1977). Reprinted by permission from Nature, Macmillan Magazines Ltd.

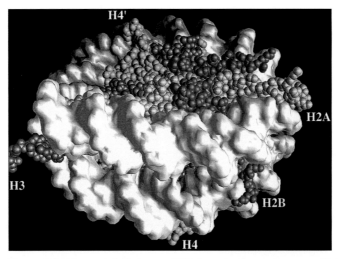

Figure 2.6. Histone tails between DNA turns in the nucleosome. Note that the tails of H3 and H2B pass the channels in the DNA superhelix (white, surface representation). Courtesy of Dr. T. J. Richmond. T. J. Richmond, Science 277: 1763 (1997).

the fact that the DNA superhelix core wrapping is not uniform but distorted by bends, this periodicity positions the minor and major grooves from neighboring turns in such a way as to line up and form channels through which histone tails can pass (Figure 2.6). This remarkable arrangement can actually enable different nucleosomes to interact and pack tightly together.

OTHER PROTEINS

With the exception of this kind of packaging in the nucleosomes, other processes of DNA require higher order nucleoprotein complexes. These are mainly replication, transcription, recombination, and transposition. For such processes to take place, DNA distortions, such as bending wrapping and looping, are necessary. Such distortions are facilitated by other sequence-nonspecific DNA-binding proteins. Some of the most important such proteins include the prokaryotic HU protein and the eukaryotic HMG (High Mobility Group). These proteins have been implicated in replication and transcription, and they are also called DNA flexers. The 3-D structure of the *Escherichia coli* HU protein and its interactions have been solved and provide a view of how these proteins bend DNA. HU is a heterodimer of two highly conserved and related subunits, creating a base consisting of two alpha helices from which antiparallel beta sheet arms are extended (Figure 2.7A). The arms contain highly charged amino acids (as in histones). HU proteins bind DNA in the minor groove in a sequence-nonspecific manner. Flexing of DNA can be achieved by interactions of the arms of the HU heterodimer as shown in Figure 2.7B.

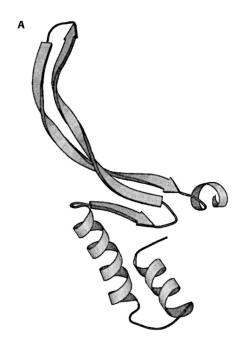

Figure 2.7. **A:** 3-D structure of *E. coli* HU. **B:** A spacefill model showing HU-DNA interactions. The two HU monomers are shown in white and red with cysteines colored yellow. G. Chaconas, Cell 85: 761–71 (1996). Reprinted with permission from Elsevier Science.

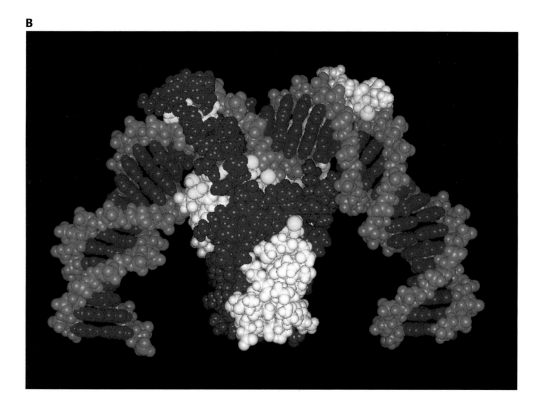

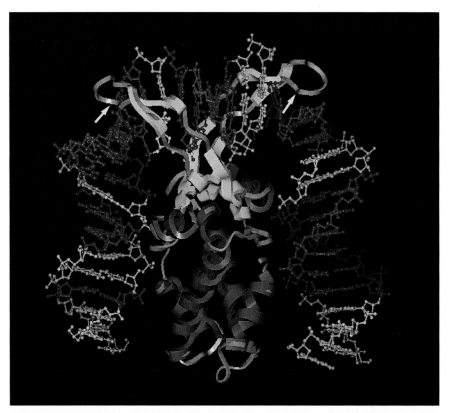

Figure 2.8. IHF-induced DNA bending. The proline intercalation resulting in the two kinks is noted by an arrow. The helices are purple, the beta sheets are cyan, and the loops are blue. A, T, G, C are red, blue, green, and yellow, respectively. P. A. Rice et al., NDB file PDT040, Cell 87: 1295–306 (1996).

Further information on the process of DNA bending has been provided by the 3-D structure of the Integration Host Factor (IHF). This small hetero-dimeric protein is necessary for building macromolecular complexes on a DNA template. Even though each subunit of IHF is nearly 40% identical to HU, it binds DNA in a sequence-specific manner. The overall structure of the subunits is very similar to HU, the body made up of alpha helices and the arms made up of beta strands. Binding of IHF to DNA bends it by 160 degrees, thus, reverting the direction of the helix axis. This bending can facilitate the interaction of other components. Its role is, therefore, architectural. Intercalation of a proline residue from both arms disrupts the base stacking and results in two large kinks that are responsible for much of the bend (Figure 2.8).

Another protein that has the ability to bend DNA but does not belong to the HU family is the Sac7d protein from the archeon *Sulfolobus acidocaldarius*. This protein binds DNA nonspecifically and bends it by a kink of 61 degrees. The kink is the result of intercalation of hydrophobic residues into the minor groove

A

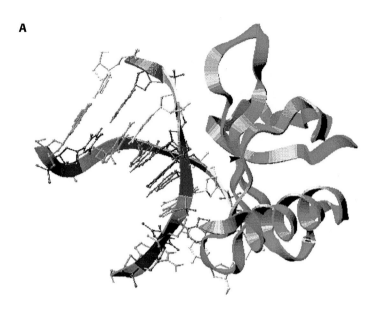

B

Figure 2.9. A: Kinking of DNA by Sac7d. Note the interface between DNA and protein characterized by the three beta-strand sheet (arrowhead) and the kinked DNA at the minor groove. S. Su et al., NDB file PD0119, J. Mol. Biol. 303: 395–403 (2000). **B:** Binding of LEF1 (Lymphoid Enhancer-binding Factor; an HMG-related protein) to the minor groove of the DNA. The binding is mediated via a three-alpha-helices fold typical for the members of this group of proteins. J. J. Love et al., PDB file 2LEF, Nature 376: 791–5 (1995).

of DNA structure. The protein itself is not distorted, due to the binding, but the DNA distortion is associated with helix unwinding around the intercalation sites and some roll between base pairs. The positioning of the intercalating amino acids is due to the alignment of a triple-stranded beta sheet. As mentioned earlier, in eukaryotes such architectural role in DNA bending is provided by the HMG proteins. In accordance with HU, IHF and Sac7d members of the HMG family bind DNA in the minor groove, but they differ in the structural motif that binds DNA. HMG proteins are characterized by a triple helix motif that confers the binding and kinking in the DNA (Figure 2.9).

CHAPTER THREE

Structure of DNA and Telomeres

——— – – –
————————————

PRIMER When we dissect chromatin, we arrive at the highest resolution, which is the DNA wrapped around histones. The structure of DNA, the double helix, needs no elaborate introduction. In this chapter, the basic elements that contribute to the 3-D structure of DNA are presented. The helical DNA can, in fact, assume different structural forms, and we discuss them along with their significance. However, not all DNA exists as double-stranded. At the ends of the chromosomes, DNA is single-stranded and is characterized by a specific nucleotide sequence. The ends of the chromosomes are called telomeres, and they can assume a distinct 3-D structure. This structure is presented along with the structure of proteins that binds telomeric sequence and plays roles in their stability that eventually affect the length of chromosomal DNA and its role in replication and cell division.

THE THREE-DIMENSIONAL STRUCTURE OF DNA

The 3-D structure of DNA is undoubtedly the most recognizable structure of a biological molecule to scientists and nonscientists alike. The determination of the 3-D structure of DNA in 1953 stunned the scientific world and completed the race for understanding how genetic information is passed on to progeny. The DNA is a simple periodic spiral structure made up of two helices. Each helix is made up of sequence of four bases connected via phosphate bonds, and the two strands are held together by a specific interaction between the bases. Adenine (A) pairs with thymine (T) through two hydrogen bonds, and guanine (G) pairs with cytosine (C) through three hydrogen bonds. Each base is part of a nucleotide (deoxynucleotide for DNA), which is the base plus a sugar plus a phosphate group. A base with sugar only is called deoxynucleoside.

The periodicity in the DNA is due to the helical turns, each turn being nearly every 10 base pairs. The arrangement of the two complementary strands in the DNA also revealed its function. Because the strands are complementary, each one can produce the other. As a result, genetic material is copied and passed to daughter cells upon division (see Chapter 4). In this sense, very few 3-D images of biological molecules reveal their function in such a spectacular way as the 3-D structure of DNA. For the two strands to form this helical structure, they must run antiparallel to each other. Let us now visualize these basic structural elements of DNA structure. First, we will examine the A-T and the C-G base pairing. A and G contain two rings, and they are called purines. T and C contain one ring, and they are called pyrimidines. The A-T and C-G pairings, therefore, deliver a structure with a standard width. In Figure 3.1, the pairs are presented indicating also the hydrogen bonding between G-C (three bonds) and A-T (two bonds). In Figure 3.1B, we can see all the components of a nucleotide (G) and the numbering of carbon atoms that determine the $5' \rightarrow 3'$ direction, which is indicated by an arrow. In Figure 3.1C, note that when the two nucleotides of the opposite strands are compared, the sugar ring is flipped to the other direction. So the direction $5' \rightarrow 3'$ is in one strand and $3' \rightarrow 5'$ in the other. This directionality is very important for proteins to recognize DNA. The sugars and the phosphate constitute the backbone of the DNA and form ridges on the edges of the helix. Between these ridges are the grooves, where the bases are exposed. But because DNA is not a regular helix, there are narrow and wide grooves, called minor and major grooves, respectively. To understand these differences, think of a perfectly symmetrical staircase. Every step of the staircase would be attached in such a way that the distance between the front and the back of each step is the same. In DNA, however, if a base pairing is a step, the distance between the sugar phosphates would be wider at one edge (major groove) and narrower at the other (minor groove) (Figure 3.2).

Let us now examine the 3-D structure of the different DNA forms in an attempt to correlate them with interactions between DNA and proteins (that will be presented in many places in this book) and with possible function. When DNA is hydrated, it exists in the so-called B-form. This structure was determined by Watson, Crick, Franklin, and Wilkins. As we can see in the 3-D images of DNA (Figure 3.3), the plane of the base pair is perpendicular to the helical axis, and the helical axis runs through the center of each base pair. The distance between every base pair is 0.34 nm. Also there is a 36-degree angle between adjacent bases on the same strand; thus, a helical turn (3.4 nm) is repeated every 10 bases. However, there are variations of these values. Such a repeating unit creates the major and the minor groove both with similar depths. The A-form of DNA is a dehydrated form (A comes from alcohol, whose addition induces dehydration) and has 11 to 12 bases per helical turn. In such a structure, the distance between adjacent bases is 0.29 nm. The plane of the base pair is not perpendicular to the helix axis but is tilted 20 degrees. Also,

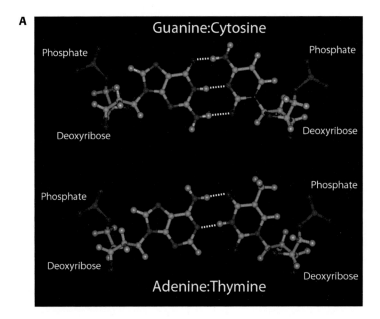

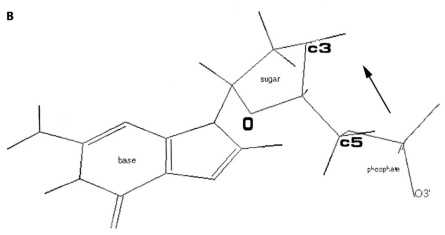

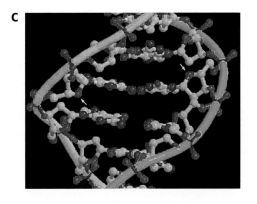

the helical axis is shifted from the center into the major groove. As a result, A-DNA has deeper major grooves and shallower minor grooves than B-DNA. A-DNA most likely does not exist in vivo, but we do know that DNA-RNA and RNA-RNA duplexes assume such a conformation. Both B- and A-DNA are right-handed. Another form of DNA is Z-DNA. This from is left-handed and is found for sequences with alternating G and C bases. This sequence is common in gene promoters and might indicate a role of such conformation in gene regulation. The Z-DNA contains 12 bases per helical turn, which covers the distance of 3.8 nm. In Z-DNA, each guanine has its sugar ring rotated 180 degrees; consequently, it bends inward to the minor groove, while the sugar ring of cytosine, as in A- and B-DNA forms, swings away from the minor groove. This sugar-phosphate backbone from a GCGCGC...sequence produces a zigzag (hence the name Z-DNA) pattern. The helix is thinner and elongated and has deep but narrow minor grooves, but the major groove does not exist.

TELOMERES

At the end of the chromosomes, DNA has a very unique sequence and structure, which constitute the telomeres. A telomere is a nucleoprotein. Specialized proteins interact with unique sequences and confer the structure and function of the telomeres. Telomeres are very important for stability of the end of the chromosomes and are made up of guanine-rich sequences, which are repeated many times. The sequence is [d(T2G4)n] in *Tetrahymena*, [d(T4G4)n] in *Oxytricha*, and [d(T2AG3)] in human. This sequence is bound by telomerase, a nucleoprotein that contains an RNA component with sequences complementary to the telomere sequences, and a protein component, which acts as reverse transcriptase. This association is very important because it does not allow the shortening of chromosomes after every cell division. During replication of linear DNA, the 5′ ends are left with the RNA primer for the last Okazaki fragment (see Chapter 4). If this gap is not filled, the ends of the DNA will become progressively shorter, and this will eventually lead to the death of the cell. However, telomerase solves this problem. The RNA component of the telomerase acts as a primer to extend the 3′ ends, thus adding telomeric sequences. After

←───

Figure 3.1. **A:** A-T and C-G pairing. Note the three hydrogen bonds between G and C and the two hydrogen bonds between A and T. The phosphate is magenta, oxygen is red, carbon is green, nitrogen is blue, and hydrogen is white. Dotted lines represent hydrogen bonding. The sugar ring is situated perpendicular to the plane of the page. Images generated by E. Fuentes. **B:** A nucleotide (G) with all its components – base, sugar, phosphate – showing the numbering of C3 and C5 in the sugar ring, which determines the direction of the strand, 5′ → 3′ (arrow). Imb-jena. **C:** A model of DNA showing the opposite direction of the sugar ring. (The arrows indicate oxygens, in two nucleotides of the opposite strands. The oxygens are colored red.) D. S. Goodsell et al., PDB file 126D, PNAS 90: 2930–4 (1993).

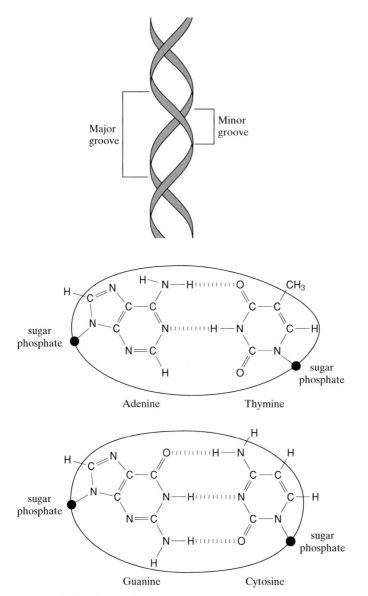

Figure 3.2. The distance between the sugar phosphates in both base pairs. On the one edge, the distance is greater than on the other. These differences constitute the structure of the major and minor groove.

removing the primer, telomeric sequences are added without net loss. The 3′ end of the telomeres is, therefore, a single-stranded overhang of the G-rich telomeric sequences. The single- or double-stranded telomeric sequences are bound by proteins, which, in fact, are responsible for the 3-D structure of the telomeres. These sequences include the two-subunit *Oxytricha nova* telomere

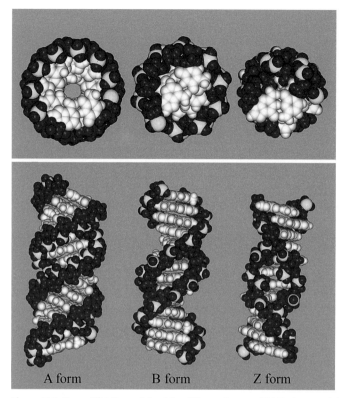

Figure 3.3. Spacefill 3-D models of the different forms of DNA along and down the axis. The bases are gray, the sugar-phosphate backbone is blue, and the phosphate is yellow. Compared with the B-DNA, note how much wider the duplex is in A-DNA and the differences in tilt of the bases. Note also the zigzag pattern of the phosphates in the Z-DNA. Images generated by E. Fuentes.

end-binding protein (OnTEBP), which binds to single-stranded telomeric sequences, and the RAP1 and TRF1, which bind double-stranded telomeric sequences.

Structure of OnTEBP and Interaction with Telomeric Sequences

OnTEBP is made up of two subunits, alpha and beta. At the alpha-beta interface, a deep cleft is created that binds ssDNA (Figure 3.4).

The protein-DNA interactions can be divided in three regions (Figure 3.5). The first includes the 3′ end sequences T8, G9, G10, G11, and G12. This sequence folds as a loop between the alpha and beta subunits of the protein. The second region includes bases G4, T5, T6, and T7. These nucleotides form an extended ssDNA-protein interface involving a stack of nucleotides and aromatic amino acids. The third region involves the 5′ end sequences G1, G2, and G3. This ssDNA-protein interface is characterized by unique folding between the ssDNA and the amino acids of the protein. Note the hydrogen bonds between G2 and Q135, which bring Y130 face to face with G2, and the hydrogen

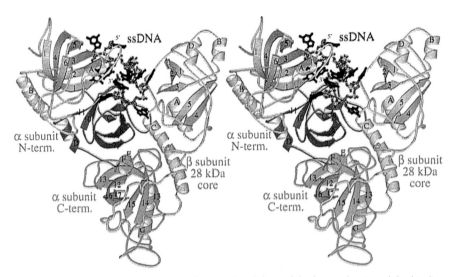

Figure 3.4. A stereo image of OnTEBP showing the alpha and the beta subunit and the binding of telomeric single-stranded DNA in the created cleft. S. C. Schultz, Cell 95: 963–74 (1998). Reprinted with permission from Elsevier Science.

bonds between G3 and K77, R272, D225, and D223. This DNA and protein fold is likely important for specificity and recognition of the telomeric sequences.

A summary of all these regions is schematically represented in Figure 3.6. Residues in alpha are in purple, residues in beta are labeled in red, hydrogen bonds are shown as dotted lines, ionic interactions appear as solid black arrows, and close contacts between hydrophobic groups are represented by orange arrows (when aromatic groups are stacked together they are indicated by an orange double-headed arrow).

Structure of Yeast RAP1

Rap1 binds the double-stranded telomeric sequences. The DNA binding domain of RAP1 consists of two defined domains and a C-terminal tail. The N-terminal domain interacts with the 3′ region of the RAP1-bindings site, whereas the C-terminal domain interacts with the 5′ region (Figure 3.7). Each domain contains a helix-turn-helix motif, which is similar to the one found in homeodomain and other DNA-binding domains (see Chapter 6).

Despite the fact that this DNA binding motif is similar to other DNA-binding motifs with specificities other than telomeric sequences, specific interactions and conformation changes in the DNA duplex may account for the recognition of conserved telomeric DNA sequences. In Figure 3.8, the interaction of basic and hydrophobic amino acids from domain 2 results in conformational change in the DNA duplex, which might be unique for recognition of telomeric sequence.

Except for these interactions between the telomeric sequences and proteins, the telomeres can assume a characteristic 3-D structure at the 3′ end overhang.

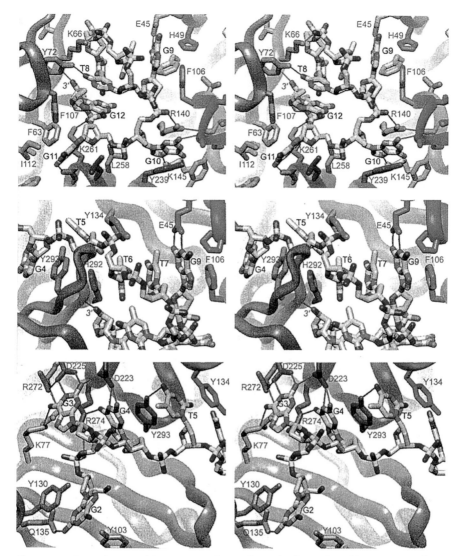

Figure 3.5. Stereo images of OnTEBP-DNA interactions. The first region of interactions is shown on the top panel, the second appears on the middle panel, and the third is displayed on the lower panel. S. C. Schultz, Cell 95: 963–74 (1998). Reprinted with permission from Elsevier Science.

This structure is called the G-quartet and is characterized by three stacked G-tetrads, which are connected by loops. This structure can be achieved in the presence of monovalent cations, but it has been also shown that OnTEBP can promote the formation of a G-quartet. To understand how this structure is formed, let us consider the human telomeric sequence d[AG3(T2AG3)3]. When this synthetic sequence was examined by NMR, the following interresidue NOEs between nonadjacent residues were found (Figure 3.9).

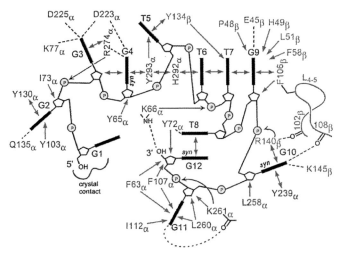

Figure 3.6. Interactions between OnTEBP with ss telomeric DNA sequences. S. C. Schultz, Cell 95: 963–74 (1998). Reprinted with permission from Elsevier Science.

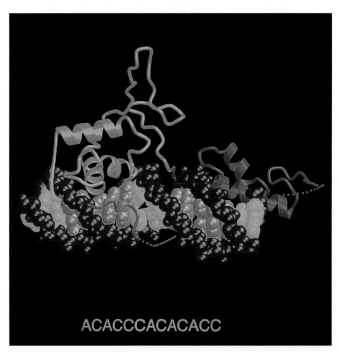

Figure 3.7. The RAP1 DNA-binding domain in complex with DNA. Domain 1 is on the left top, and domain 2 is on the right. Note that the C-terminal tail (arrowhead) emerges from domain 2, folds back toward domain 1, and interacts with the major groove. The bound DNA sequences are also indicated. D. Rhodes, Cell 85: 125–36 (1996). Reprinted with permission from Elsevier Science.

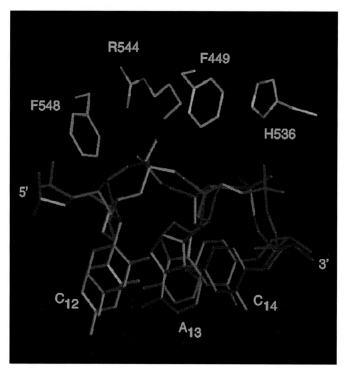

Figure 3.8. Interaction of domain 2 amino acids with C12-A13-C14 (blue). A B-DNA model (magenta) is superimposed, allowing the demonstration of the changes in DNA conformation that could be unique for telomeric sequences. D. Rhodes, Cell 85: 125–36 (1996). Reprinted with permission from Elsevier Science.

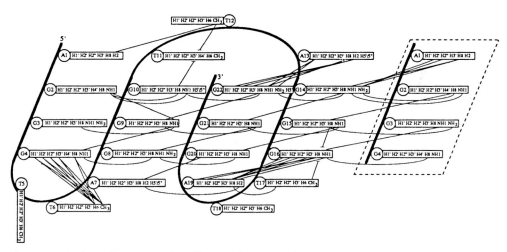

Figure 3.9. A schematic representation of interresidue contacts between nonadjacent residues in the human telomeric sequence. Note the interactions between G2G10G22G14, G3G9G21G15, and G4G8G20G10, which lead in the formation of three G-tetrads. T6 interacts with G4, A1 interacts with G14, and A19 interacts with G16; these interactions identify loops over adjacent tetrads. D. J. Patel, Structure 263–82 (Dec 15 1993). Reprinted with permission from Elsevier Science.

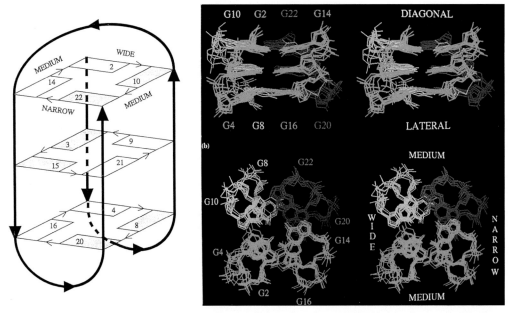

Figure 3.10. On the left, a schematic diagram of the G-quartet is shown. The backbone tracing is shown by a thick line, and the direction of the chain is represented by thick arrows. Guanines are represented by rectangles and numbered by residue position. On the right, stereo images of the three tetrads are presented. The bottom image is a view looking down the helix axis. The three different groove widths are labeled in both as wide, medium, and narrow. The positions of the diagonal and lateral loops are also labeled. D. J. Patel, Structure 263–82 (Dec 15 1993). Reprinted with permission from Elsevier Science.

The folding of the previous representation at a 3-D space would identify the characteristic structure of the G-quartet (Figure 3.10).

However, the ends of the chromosomes are not linear. Another factor that binds double-stranded telomeric sequences, TRF2 can remodel linear telomeric DNA into large duplex loops, called t-loops. These t-loops are created when the

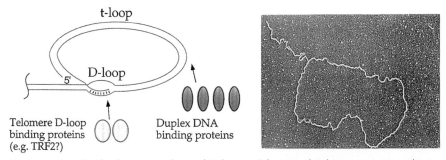

Figure 3.11. Left: The formation of t- and D-loops. Right: t- and D-loops as seen under an electron microscope. C. W. Greider, Cell 97: 419–22 (1999) and J. D. Griffith, Cell 97: 503–14 (1999). Reprinted with permission from Elsevier Science.

3′ telomeric overhang invades the duplex telomeric repeat. A D-loop is created between the 3′ overhang; sequences of the neighboring telomeric sequence of the same strand are shown in Figure 3.11. According to this arrangement, we could argue that a t-loop might protect the end of the chromosome from cellular activities that can act on the ends of linear DNA, such as activation of DNA checkpoints. Also, it might help cells distinguish naturally broken DNA from chromosomal ends.

CHAPTER FOUR

DNA Replication

——— – – –

PRIMER DNA replication involves many proteins acting at different steps. DNA polymerases are the main enzymes needed to create new strands from the parental ones. However, DNA polymerase alone cannot complete the whole task of replication. When the double-stranded DNA melts at the origin of replication to expose the single strands to the polymerizing enzyme, the fork must continue to unwind until the termination signals have been reached. This part in replication is performed by helicases, and the topology ahead of the fork (mainly coiling and superwinding of the DNA) is controlled by topoisomerases. Finally, termination is achieved by the interaction of both helicases and termination proteins. Editing is another important function during replication to ensure that misincorporations are eliminated and, thus, mutations are avoided. Also, other proteins are involved in securing the processivity of DNA polymerase and the processing of the Okazaki fragments. The 3-D structure of all the major players participating in the entire process of replication is for the most part known. For some of them, we also know how they interact with each other or with DNA. This enables us to visualize replication in a 3-D journey. This chapter introduces the 3-D structure of DNA polymerases, the mechanism of incorporation of the incoming nucleotide, and how DNA polymerase interacts with DNA and other proteins to achieve processivity and editing. We then examine the structure of different helicases, their interaction with DNA, and the mechanism of unwinding as suggested by the 3-D structures. Similarly, the function of topoisomerases and termination proteins will be revealed at the 3-D level. Because some of the players in replication are not the same in both prokaryotes and eukaryotes, when appropriate we present them separately. In all, we will view the whole process of replication as it unfolds at the 3-D level, which provides a unique and stunning view of the mechanisms involved.

When a cell divides to produce two daughter cells, the correct passing of genetic information to both of them becomes a very crucial event. For this, the DNA should be faithfully duplicated so that both cells end up with exactly the same genetic blueprint. The structure of DNA itself can explain how this occurs. The two strands of DNA dissociate at a particular site and become templates for the synthesis of the complementary strands. However simple this might sound, replication involves some very complicated events, and we still do not know all the mechanisms involved in the different stages from initiation to termination. Replication starts at the so-called origins of replication. These origins are sites in the genome where the DNA strands separate and replication starts. At this point, the process of elongation begins with the synthesis of the new strands. The main player in this event is DNA polymerase, which incorporates the nucleotides into the growing strand. During this process, the DNA should be continuously unwinding, and this is possible by the action of helicases. At the same time that the double-stranded DNA opens, downstream stress can result in compacted DNA or knotting that will inhibit replication. Other enzymes, the topoisomerases, solve this problem (Figure 4.1). Finally, replication stops at specific sites with the help of termination-specific proteins. In the next pages,

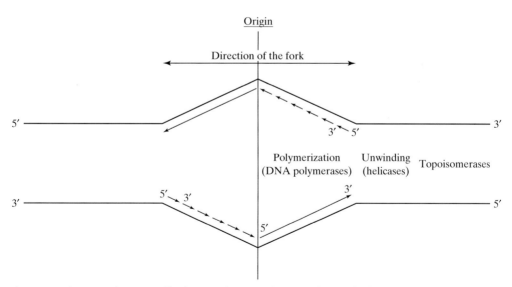

Figure 4.1. The general process of bi-directional DNA replication indicating the level of action of the main enzymes involved.

we will examine all these events in detail, and several mechanisms of DNA replication will be revealed at the 3-D level.

THE PREREPLICATIVE COMPLEX

As already mentioned, DNA replication begins at specific sites called origins of replication. These sites contain consensus sequences that are recognized by proteins that aid in DNA melting, which is necessary for the recruitment of DNA polymerase and the initiation of replication. These origins and the factors involved are not the same in bacteria and eukaryotes. Therefore, we will examine replication origins separately.

Replication Origins in Prokaryotes

The bacterial genome is circular and small; therefore, there is only one origin of replication. DNA replication proceeds bi-directionally from the origin. If, for example, replication starts at 12 o'clock, it proceeds both ways and it terminates at 6 o'clock. However, exceptions to the bi-directionality of replication do exist, with one example being the plasmid ColE1. In *E. coli*, the origin site is called oriC and is nearly 250 bp long. OriC contains some very crucial consensus sequences. There are four 9-mers with the consensus sequence TTATCCACA. Two of them are in one orientation, and the others are in the opposite orientation. Also, there are three 13-mers, all of which appear as a continuous segment characterized by the sequence GATC, which is the BglII site. The initial complex that binds to these sequences consists of HU protein (DNA flexer, see Chapter 2) and Dna:ATP. First, these two proteins bind the previously mentioned 9-mer consensus sequences. The bending introduced by HU destabilizes the 13-mer repeat and induces its melting. This reaction allows protein DnaB to bind to the single-stranded region. DnaB is helped in this by DnaC (Figure 4.2). After this stage, where the prepriming complex has been formed, a primase binds, and the primosome can now synthesize primers to initiate DNA synthesis. Certain circular DNAs, especially the ones found in phages, replicate by a rolling circle mechanism. During this type of replication, the double-stranded DNA is nicked, and the 3' end is extended and uses the intact circular strand as a template. Because the 5' end is displaced continuously, the new strand is synthesized as a circle.

Replication Origins in Eukaryotes

The eukaryotic genome, unlike the prokaryotic, is linear and much longer. Consequently, there are many origins of replication. In yeast, we can find a replication origin every 40 kbp, and they are contained within the so-called autonomously replicating sequences (ARSs). ARSs are composed of four important regions: A, B1, B2, and B3. Analysis from mutations along the ARS showed that region A is the most important because mutations in that region abolish all ARS activity. The A region, which is 15 bp long, is characterized

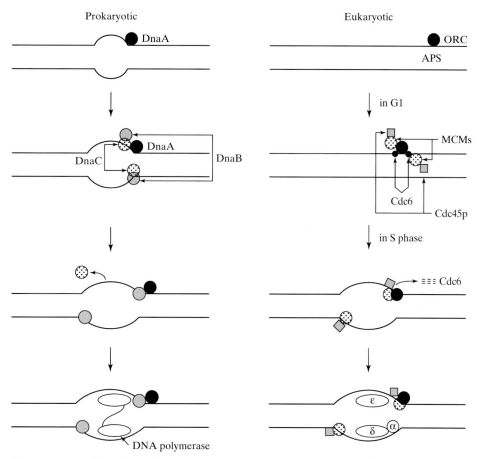

Figure 4.2. Assembly at the replication origins in prokaryotes and eukaryotes. Four distinct steps represent the initiator binding (recognition of the origin of replication), recruitment, remodeling, and polymerase loading.

by an 11-mer consensus sequence T/ATTTAT/GA/GTTTT/A, which seems to be an inverted repeat. The other regions have a lesser effect, with B3 involved in DNA bending. The consensus sequence of region A acts as a core for the origin recognition complex (ORC), a six-polypeptide complex, which appears to be analogous to the prokaryotic DnaA. At the end of mitosis, ORC is bound at the origin. During G1 the minichromosome maintenance (MCM) proteins are loaded on the ORC with the help of the kinase Cdc6p. Another kinase, the Cdc45p, is also needed in the prereplicative complex. In the S phase and as replication is initiated, Cdc6p is degraded, and DNA polymerase is recruited. The other members of the complex, MCM and Cdc45p seem to track along the replication fork during the elongation process (Figure 4.2). Members of the MCM family contain motifs that are characteristic in DNA-dependent ATPases, and they act as helicases. Replication in eukaryotes also proceeds bi-directionally.

ELONGATION AND THE SYNTHESIS OF NEW DNA STRANDS

As previously mentioned, synthesis of new DNA strands is catalyzed by DNA polymerase. This enzyme is very interesting and has several functions. First, it can synthesize DNA from the $5' \rightarrow 3'$ direction. Figure 4.1 immediately tell us that synthesis of both strands couldn't occur the same way. One of the new strands, which uses the $3' \rightarrow 5'$ parental strand as template, would be synthesized without any problem, and its synthesis would be continuous and along the direction of fork movement. This is called the leading strand. However, the same continuous synthesis cannot happen using the parental $5' \rightarrow 3'$ strand because a continuous synthesis as the leading strand would require a DNA polymerase that can act from $3' \rightarrow 5'$. This lagging strand problem is solved by synthesizing small DNA fragments, called Okazaki fragments, in the correct $5' \rightarrow 3'$ direction (opposite to the fork movement). These fragments are joined together and eventually lead to the complete synthesis of the lagging strand. This difference in synthesis between leading and lagging strands also implies the involvement of other enzymes, among them primase that synthesizes the RNA primers for the initiation of Okazaki fragment synthesis and ligase, which joins the Okazaki fragments.

DNA polymerase, however, has other functions as well. A very important one is its $3' \rightarrow 5'$ exonuclease activity. In other words, this enzyme can go the opposite direction of DNA synthesis and excise the $3'$ nucleotides. This capability helps in repair, when a mismatch has occurred. It also has a $5' \rightarrow 3'$ activity, which aids in removal of RNA primers for the Okazaki fragments. Therefore, because of these different functions of DNA polymerase, there are, in fact, several DNA polymerases. Evolutionarily speaking, the different DNA polymerases are related, they differ in function and structure.

DNA Polymerases

In prokaryotes there are three DNA polymerases. DNA polymerase I can act in replication repair and removal of RNA primers. In other words, it has a $5' \rightarrow 3'$ polymerase activity, a $3' \rightarrow 5'$ exonuclease activity used for mismatch correction and a $5' \rightarrow 3'$ exonuclease activity used for the removal of RNA primers. With trypsin digestion, the $5' \rightarrow 3'$ polymerase domain and the $3' \rightarrow 5'$ exonuclease domain can be isolated and is called the Klenow fragment. This enzyme is mostly used when scientists prepare probes in the laboratory. DNA polymerase II lacks the $5' \rightarrow 3'$ exonuclease activity; therefore, it cannot be used for RNA primer removal. Its main function is in DNA repair. DNA polymerase III also lacks the $5' \rightarrow 3'$ exonuclease activity, but it is the main DNA polymerase used for de novo DNA synthesis as occurs in chromosome replication. In other words, DNA pol III is the only polymerase required for replication.

In eukaryotes, depending on the species, there are different DNA polymerases. For example, there are five in mammals and three in yeast. DNA polymerase α plays a role in DNA replication and synthesis of the lagging strand. In fact, this enzyme is the only one that has primase activity. DNA polymerase δ plays a role in the synthesis of the leading strand. DNA polymerase ε is similar to δ but is most likely involved in repair due to its $3' \to 5'$ exonuclease activity. DNA polymerase β is also involved in repair. Finally, DNA polymerase γ is located in the mitochondria and also has a $3' \to 5'$ exonuclease activity. It is, therefore, involved in replication of mitochondrial DNA.

A very important characteristic of DNA polymerases is their processivity. By processivity, we mean the ability of DNA polymerase to bind the DNA strand well and not fall easily after replication starts. Processivity enables the enzyme to synthesize DNA for a long time. Replication occurs at a speed of 1,000 nucleotides per second, and if the enzyme is not processive, it cannot replicate DNA efficiently. In bacteria, DNA pol III is highly processive, and, as we will see, it owes its processivity to one of its domains that binds DNA very tightly. The analogous protein in eukaryotes is the proliferating cell nuclear antigen (PCNA), which also works as a clamp and enhances the processivity of DNA polymerase delta by a factor of 40. None of the eukaryotic polymerases has a $5' \to 3'$ exonuclease activity for the removal of the RNA primers. This activity is provided by FEN-1 ($5' \to$ exonuclease 1 of flap endonuclease 1), which also interacts with PCNA. We will examine in detail the 3-D structure of the clamps and the interactions with DNA and FEN-1 in the next sections.

First, we will proceed to examine the three-dimensional structure of DNA polymerase in order to probe the mechanisms involved in DNA replication.

The Three-Dimensional Structure of DNA pol I

Most structural studies have been undertaken with the Klenow fragment. This is enough to provide us with the important features involved in DNA synthesis and repair. The 3-D structure of the prokaryotic Klenow fragment has been solved from many sources including *E. coli, Bacillus stearothermophilus, Thermus aquaticus*, and bacteriophage T7. The structure reveals that the enzyme is made up of a large domain and a small domain. The large domain has the polymerase activity, and the small has the exonuclease activity. The large domain resembles a right hand, so the major features of this part are designated as fingers, palm, and thumb. The most prevalent feature of this domain is a cleft 20 to 24 Å wide, big enough to bind B-DNA. The palm is basically made of beta sheet and is the place where the double-stranded DNA lies in a groove created by the fingers and the thumb. The fingers are composed of a number of helices that form a wall 50 Å long and create the center of catalysis, where the incoming nucleotide is incorporated in the growing chain. The other wall of the cleft is formed mainly by two alpha helices (thumb) that hold the double-stranded DNA (Figure 4.3). The DNA is in the B-form except for the last two base pairs at the $3'$ end of the primer where it adopts the A-form. A sharp turn exposes the template to the incoming nucleotide.

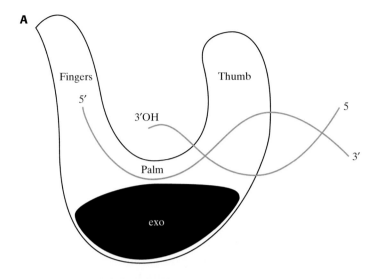

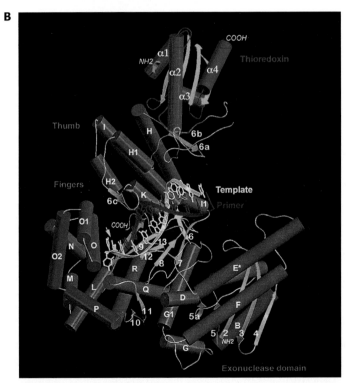

Figure 4.3. A: Schematic illustration of the Klenow fragment showing the arrangement of the fingers, palm, thumb, exonuclease, and DNA in the active site. Upon mismatch, the 3′-OH falls into the exo domain. **B:** 3-D structure of the T7 DNA polymerase with the large domain featuring the fingers, palm, and thumb. The small domain has the exonuclease activity. Thioredoxin is recruited from the host *E. coli* as processivity factor. The incoming nucleotide is cyan. T. Ellenberger, Nature 391: 251–8 (1998). B reprinted by permission from Nature, Macmillan Magazines Ltd.

The Incorporation of the Incoming Nucleotide

Let us now visualize how the enzyme captures the incoming nucleotide and incorporates it in the growing DNA chain. To be able to catch the polymerase in action, a catalytically active enzyme should be crystallized. In other words, as the reaction of DNA replication occurs, a sample containing the enzyme, template, primer, and free nucleotides should be crystallized. This procedure can enable the experimenter to obtain snapshots of the action. Such an experiment has virtually revealed the mechanism of incorporation. The structure that we will explore is the T7 DNA polymerase with bound template and primer in a buffer containing dideoxy nucleotides. When a dideoxy nucleotide is incorporated, it inhibits incorporation of other nucleotides. This is an important experiment in which to capture the moment of incorporation. For the incorporation to occur, DNA polymerases catalyze phosphoryl transfer by a nucleophilic attack. The attacking part is the 3'-OH group of the primer on the alpha-phosphate of the incoming nucleotide and results in incorporation of the nucleotide and release of pyrophosphate. Remember that the free nucleotides have three phosphates (alpha, beta, gamma), but the incorporated ones have only the alpha.

Several sequence motifs that are conserved in different DNA polymerases have been implicated in this reaction. Among them, residues of the O helix are very important. In Figure 4.4, we can explore some of these interactions. The incoming nucleotide stacks between the O helix (517–530) and the DNA at the primer's 3' end. The template cytosine that will form a base pair has its 5' and 3' phosphates hydrogen-bonded with Gly-533 (of helix O) and His-607 (of helix Q), respectively. Its pyrimidine ring stacks against the conserved Gly-527 of helix O (Figure 4.4).

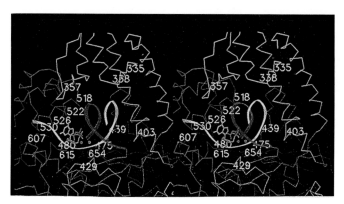

Figure 4.4. A stereo image indicating the positions of the incoming nucleotide (gray), the template (yellow), and the primer (magenta) in the active site of T7 DNA polymerase. The two gray dots are metal ions. T. Ellenberger, Nature 391: 251–8 (1998). Reprinted by permission from Nature, Macmillan Magazines Ltd.

Let us now look closely at the active site and the interactions of the incoming nucleotide (ddGTP). As seen in Figure 4.4, the incoming nucleotide stacks between helix O and the 3′ end of the primer. However, the 3-D structure showed that the incoming nucleotide is positioned in the right place by interactions with conserved residues and extensive contacts of all three phosphates with two metals. The two metals interact with two conserved aspartic acids, Asp-475 (of palm's beta stand 9) and Asp-654 (of palm's beta strand 13). These interactions are aided by two water molecules. Another residue, Ala-476 (of palm's beta strand 9) is also involved in these interactions (Figure 4.5). Also, two oxygens of the gamma phosphate of the incoming nucleotide are contacted by Arg-518 and Tyr-526 (helix O). The beta-phosphate is contacted by His-506 (helix N), and the alpha phosphate, by Lys-522 (helix O). Overall, we can see that helix O makes extensive contacts with the incoming nucleotide, and that the template is largely contacted by the palm region. This three-dimensional snapshot clearly captures the very event of DNA synthesis in action. The catalytic site of T7 DNA polymerase is structured in such a unique way and, with the involvement of the two metal ions (Zn^{2+} and Mn^{2+}), is able to guide the incoming nucleotide to the correct place for polymerization. Obviously, the metals, except for orienting the incoming nucleotide, help in the phosphoryl transfer reaction, with metal A suitably located to activate the 3′-hydroxyl of the primer (Figure 4.5). These metals have been found in other divergent

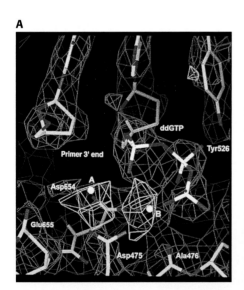
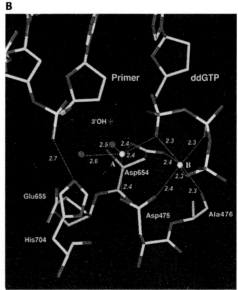

Figure 4.5. A: A close-up of the T7 polymerase catalytic site. Note that the incoming ddGTP is stacked between the 3′ end of the primer and Tyr-526 of helix O. Two metals (A and B) interact with Asp-475 and Asp-654. **B:** The metals ligate to all phosphates of ddGTP with the help of two water molecules (red spheres). Note how close metal A is to the primer's 3′-OH (cross). T. Ellenberger, Nature 391: 251–8 (1998). Reprinted by permission from Nature, Macmillan Magazines Ltd.

polymerases, such as the mammalian DNA polymerase-beta, and this shows how important this mechanism is for DNA polymerization.

Let us now focus again at the catalytic site to see the fitting of the incoming nucleotide in relation to misincorporation and proofreading. Figure 4.6 presents a space-filling model of the catalytic site with the primer-template bases and

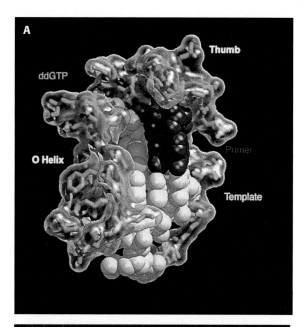

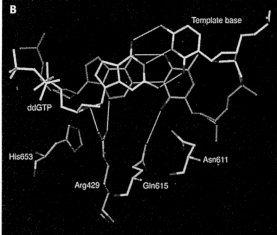

Figure 4.6. **A:** Spacefill model showing the site where the incoming ddGTP (cyan) pairs with a cytosine of the template (yellow) and stacks against primer's 3′ adenine (magenta) and the O helix. The two metals are shown as red spheres. **B:** The nascent base pair (G-C; white) in the polymerase active site and the adjacent base pair of the primer-template (A-T; orange). Note the interactions between residues Arg-429 and Gln-615 with the minor groove of the A-T pair. T. Ellenberger, Nature 391: 251–8 (1998). Reprinted by permission from Nature, Macmillan Magazines Ltd.

the incoming nucleotide ddGTP. ddGTP (cyan) stacks against the adenine at the 3′ end of the primer (magenta) and forms a base pairing with a cytosine (yellow). The nascent base pair lies in a narrow slot that obviously does not favor mismatched or staggered base pairs (Figure 4.6A). Also, the template-primer base pair adjacent to the nascent one is contacted by the conserved residues Gln-615 of helix Q and Arg-429. The resulting hydrogen bonds are in the minor groove (Figure 4.6B). A mismatch at this position could result in the loss of these interactions in the minor groove because DNA has a unique structure, with a standard width that would be altered if a mismatch occurs. After a mismatch has been sensed, the mismatched portion is translocated in the 3′ → 5′ exonuclease domain for excision.

The structure of the polymerase domain (as seen in T7 DNA polymerase) is basically the same in other polymerases as well. They all have the characteristic U shape with fingers, palm, and thumb and similarly positioned conserved catalytic residues. At this point, we do not need to examine the structure of other DNA polymerase I; however, it will be useful to examine the structure of some eukaryotic DNA polymerases and compare them with prokaryotic DNA polymerase I.

The Three-Dimensional Structure of the DNA Polymerase from Phage RB69

This polymerase (gp43) is a member of the eukaryotic DNA polymerase alpha family. Its polymerase domain has the characteristic U shape with fingers, palm, and thumb; however, there are some marked differences. The fingers and thumb have different topologies than those seen in DNA polymerase I. The fingers are made mainly of two antiparallel alpha helices; but the palm is topologically similar to the one in DNA polymerase I. The only difference is that gp43 has one of the three carboxylates (the amino acids that bind the incoming nucleotide) on a different beta strand. The overall structure of gp43 with the polymerase domain and the N-terminus domain, which contains the exonuclease domain, has a disk structure with a hole in the middle. When we take the N-terminus domain into consideration, another difference between gp43 and DNA polymerase I is obvious. The editing exonuclease domains lie on opposite sides (Figure 4.7). When RB69 DNA polymerase was crystallized with primer template DNA and dTTP, common mechanisms between this polymerase and pol I were suggested. The finger movement in relation to the apo-protein was similar as were the minor groove interactions near the primer 3′ end.

The Three-Dimensional Structure of Human DNA Polymerase Beta

The main job of human DNA polymerase beta is to fill single nucleotide gaps in DNA produced by the base excision repair (BER) pathway. The enzyme is composed of only two domains. The 8-kDa N-terminal domain has the

A

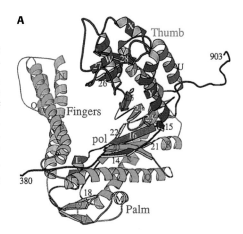

Figure 4.7. A: The 3-D structure of the gp43 polymerase domain showing the over-all U shape and the location of fingers, palm, and thumb. Compare to T7 DNA polymerase (Figure 4.3). B: Surface representations of RB69 gp43 and the Klenow fragment (KF). Note the difference in the location of the exo-nuclease domain compared to that of T7 DNA polymerase. T. A. Steitz, Cell 89: 1087–99 (1997). Reprinted with permission from Elsevier Science.

B

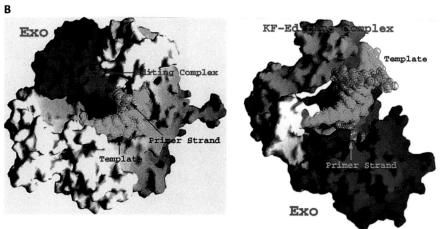

deoxyribose phosphate lyase activity (excision of remaining deoxyribose phos-phate residue; in BER the damaged base is hydrolyzed by a DNA glycosylase, the backbone is incised by an AP-endonuclease, and then the phosphate is excised). The C-terminal domain has the nucleotidyl transfer activity, which is responsible for filling the gap with an incoming nucleotide. This domain con-tains the characteristic U shape ensemble with fingers, palm, and thumb. The catalytic site of this enzyme is very similar to the one seen in other DNA poly-merases. The 3'-OH of the primer makes an in-line nucleophilic attack on the alpha phosphate of the incoming nucleotide. The alpha phosphate is coordi-nated by two metals that are anchored to the active site by three conserved aspartic acids (Figure 4.8).

When the structure of the polymerase complexed with gapped DNA (binary complex) is compared with the same structure including ddCTP (ternary complex), some interesting features are revealed. Upon ddCTP binding, the thumb subdomain rotates, producing a closed conformation to contact the

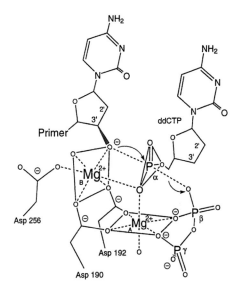

Figure 4.8. Nucleotidyl transfer mechanism of polymerase beta. Note the attacking 3′-OH end of the primer, the incoming nucleotide (here is a ddCTP), and the coordination with the Mg^{2+} metal ions and aspartic acids. M. Sawaya, Biochemistry 36: 11205–15 (1997). Reprinted with permission from American Chemical Society.

ddCTP-template base pair. This movement results in further conformation changes, which poise the catalytic residue Asp-192 (see Figure 4.9), dNTP, and template for nucleotidyl transfer, thus assembling the active catalytic site (Figure 4.9A). Such a movement provides evidence for an induced fit mechanism. According to this mechanism, the conformational changes observed after binding of the correct nucleotide would result in proper alignment of the catalytic groups. Such an alignment cannot be achieved for a nucleotide that is not complementary to the template. Such movement has also been seen in T7 DNA polymerase; however, in this case, the O helix moves, creating an open and closed conformation of the fingers. Also, the polymerase binds gapped or nicked DNA with a 90-degrees kink. This kink occurs at the 5′-phosphodiester linkage of the template nucleotide (Figure 4.9A) and enables the contact between the thumb and the incoming dNTP-template base pair, which is important for the checking mechanism.

Editing

When a mismatch occurs in the polymerase active site, several sensing mechanisms result in the displacement of the 3′ terminus of the primer to the exonuclease domain for excision. In both pol I and pol alpha families, the polymerase active site and the editing exonuclease site are 30 to 40 Å apart. Therefore, the 3′ primer end must travel a considerable distance. The three-dimensional structures of DNA polymerases complexed with DNA at both sites have revealed some interesting conformational changes in both the thumb and the trajectory of the DNA. When DNA is in the editing site, part of the thumb is displaced, allowing the single-stranded portion of the primer template to enter the active site of the exonuclease domain. Also, the DNA is positioned

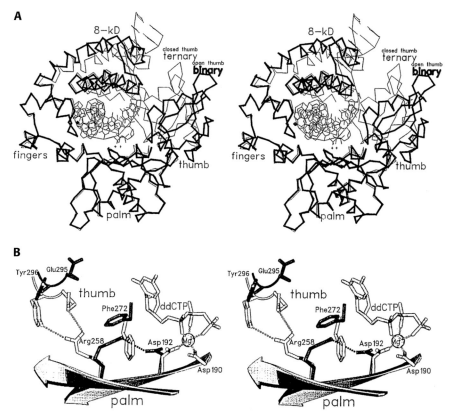

Figure 4.9. A: Stereo picture depicting the thumb movement upon ddCTP binding. The thick lines represent the structure of the binary complex, and the thin lines denote the structure of the ternary complex. Note the change in thumb conformation and to a lesser degree the change in the 8-kDa domain. Also note the kink in the DNA, the incoming ddCTP, and the two metals (the two crosses). **B:** The changes in the catalytic site in response to the thumb movement. The binary complex is black, and the ternary is gray. Thumb closure assembles the active site by moving Phe-272 in position capable of disrupting the Arg-258–Asp-192 hydrogen bond. This frees Asp-192, which in turn binds the two metal ions. Also Glu-295 and Tyr-296 form hydrogen bonds with Arg-258 to prevent any interference with the Asp-192 conformation. M. Sawaya, Biochemistry 36: 11205–15 (1997). Reprinted with permission from American Chemical Society.

differently in the two active sites. Going from the polymerase site to the editing site, the DNA trajectory twists about 30 degrees (Figure 4.10). Such changes are common in most polymerases studied to date. In the Klenow fragment, it has been shown that four nucleotides of single-stranded DNA from the primer are bound in the editing site but that three nucleotides are needed in the RB69 DNA polymerase.

Processivity of DNA Polymerases and the Structure of the Clamp

As indicated previously, DNA polymerase should be able to stick to the DNA for a long time to polymerize thousands of nucleotides. If a polymerase

A

B

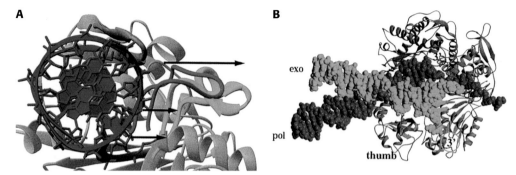

exo

pol

thumb

3′

Figure 4.10. A: RB69 DNA polymerase (gray) complexed with DNA (red), looking down the helical axis. A portion of the thumb (dark blue in the polymerase site) is displaced (cyan) when DNA is in the editing site. **B:** Comparison of the primer template bound in the polymerase (red) or the exonuclease site (blue). Note the flipping of the 3′ end from the one site to another. A particular beta hairpin (colored green) of the exonuclease domain becomes a wall between the two active sites so that the 3′ end cannot go back to the polymerase site before editing. This hairpin is colored gold in A. T. A. Steitz, Cell 99: 155–66 (1999). Reprinted with permission from Elsevier Science.

is not able to do this and falls off the DNA strand, the polymerase is not processive and cannot be used for replication. In prokaryotes, for example, DNA polymerase I is not processive and is used mostly for repair and degradation of RNA primers. Instead, DNA polymerase III is highly processive and is used for de novo synthesis of DNA. DNA polymerase III owes its processivity to its complex structure (when compared with pol I), which contains, among other subunits, one that clamps to the DNA strand and allows it to slide for long distances. To understand the function of the clamp, we should first familiarize ourselves with the overall structure of DNA polymerase III.

The most basic unit of DNA polymerase III is its catalytic core. This is composed of subunits α, ε, and θ. This core is able to synthesize DNA, but it lacks processivity. Subunit α is equivalent to the polymerase domain seen in the other polymerases described earlier, having the characteristic U shape containing the fingers, palm, and the thumb. Subunit ε has the $3' \rightarrow 5'$ proofreading activity, and subunit θ most likely is used for assembly of the catalytic core. Two catalytic cores with two τ subunits create pol III*, which has increased processivity. pol III′ consists of pol III* and the gamma/delta (γ/δ) complex. The additional subunits in pol III′ are γ, δ, δ', χ, and ψ. Subunit γ is the catalytic site of the clamp loader and is homologous to δ' (see below). Subunits χ and ψ contribute to the stability of the complex. The holoenzyme consists of pol III′ plus two β subunits, which clamp the enzyme to DNA, and is responsible for synthesis of the leading and lagging strands. Two beta subunits create a ring that surrounds the DNA duplex (Figure 4.11).

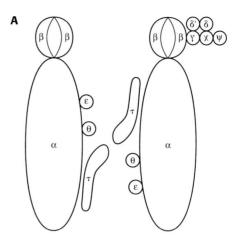

Figure 4.11. A: The organization of the different subunits in DNA polymerase III. B: Two β subunits creating a ring structure that binds to the DNA strands. This is the clamp or processivity factor that allows DNA polymerase III to bind the DNA tightly and synthesize long strands. J. Kuriyan, Cell 69: 425–37 (1992). B reprinted and with permission from Elsevier Science.

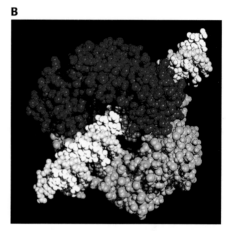

The Loading of the Clamp

The Pol III clamp loader is the γ/δ complex. The primer template DNA seems to load the β subunit as follows. First, the complex binds ATP and undergoes a conformation change. This change exposes the β-binding surface of the complex's δ subunit. Binding subunit β to δ induces an opening in the β ring. Interactions of β and δ subunits are prevented by interaction of δ with δ'. Then the γ complex recognizes the primer template DNA and brings β to the DNA. Next, ATP (adenosine triphosphate) hydrolysis or ADP (adenosine diphosphate) release destabilizes the β-δ interaction, resulting in closing the β ring around the DNA. Finally, the polymerase joins the β clamp and proceeds to polymerize DNA. After completion of an Okazaki fragment, the core polymerase is separated from the β subunit. The clamp can now slide to the next primer. Also, it seems that the lagging-strand polymerase can transfer between different clamps, without separating from the complex, even before the Okazaki fragment synthesis is finished because there are more beta subunits than polymerase assemblies.

Processivity Factors of Eukaryotic DNA Polymerases

Unlike DNA pol III, the clamp in eukaryotic DNA polymerases is trimeric. We do have structural information of the RB69 and human DNA polymerase δ clamp. For human DNA pol δ and ε, the clamp is the proliferating cell nuclear antigen. This trimeric unit is ring shaped and surrounds DNA. The interaction of these clamps with DNA polymerase has also been studied at the three-dimensional level, providing details of their association. The C-terminus of RB69 DNA polymerase has the sequence KKASLFDMF. This peptide has been found to interact with the clamp. This interaction is very similar to the interaction of PCNA with the similar peptide RQTSMTDFY found in the cell cycle check point protein p21(CIP1). This protein competes with DNA polymerase and arrests DNA replication. Therefore, p21(CIP1) must interact with a domain in PCNA and mask it from interacting with DNA polymerase (Figure 4.12). The clamp loader in eukaryotes is the cdc6 protein that we encountered in the origin of replication. For polymerase γ, the processivity factor is the accessory subunit polγB, which functions as a homodimer. The processivity factor appears to be similar to glycyl-tRNA synthetase and dimerizes through a four-helix bundle. In mutants without the four-helix bundle, the factor failed to dimerize, but it did bind to the catalytic subunit A. This response indicates that the functional holoenzyme contains the catalytic subunit A and two polγB molecules.

Synthesis of the Primers in the Lagging Strand

As indicated in the beginning of this chapter, for DNA synthesis in the lagging strand, an RNA primer should be synthesized before initiating Okazaki fragments. Synthesis of these primers is accomplished by a specialized RNA polymerase, the primase. The *E. coli* primase (DnaG) is known to interact with the DnaB helicase, which unwinds the DNA strands at the fork (see later), the single-stranded DNA-binding proteins (SSB), and the DNA polymerase III holoenzyme. DnaG is capable of synthesizing 60-nucleotide-long primers in vitro, but in the replisome the length is about 11 nucleotides. Based on sequence analysis, prokaryotic primases seem to be distinct from the ones found in archaea and eukaryotes. DnaG is composed of three structural domains. The N-terminal domain, which is the zinc-binding domain called ZBD; the core domain, which contains the polymerase region; and the C-terminal domain, which is called DnaBID and interacts with the helicase DnaB.

The three-dimensional structure of the DnaG core region is known and has shed some light in the function of the enzyme. The core region is made up of three subdomains (Figure 4.13). The central subdomain (colored blue) is characterized by a structure of five-stranded beta sheet surrounded by six alpha helices. This structure belongs to the so-called toprim fold family seen in metal-binding phosphotransfer proteins including nucleases and topoisomerases. The

A

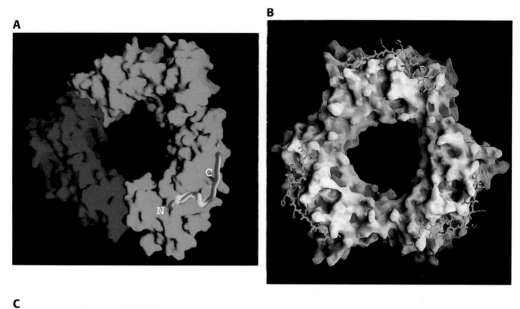

C

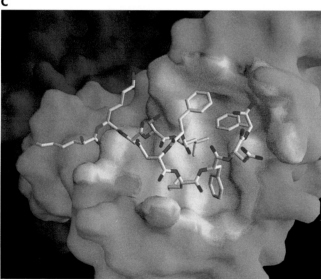

Figure 4.12. **A:** The trimeric RB69 DNA sliding clamp and the interaction with the C-terminus of RB69 DNA polymerase (yellow). T. A. Steitz, Cell 99: 155–66 (1999). **B:** Structure of the PCNA and interactions with the p21(CIP1) peptide. Note the similarity between the two clamps and also the similarity of the interaction with the different peptides. In A the p21(CIP1) peptide (red) is superimposed with the C-terminus peptide of RB69 to show this remarkable similarity. J. Kuriyan, Cell 87: 297–306 (1996). **C:** Close-up of the interaction between the C-terminus of RB69 DNA polymerase and the sliding clump, showing that it is mostly close fitting in a pocket and consists of hydrophobic interactions. Blue is one of the trimeric subunits, and yellow is the surface within 4.5 Å of the C-terminus of DNA polymerase. T. A. Steitz, Cell 99: 155–66 (1999). A, B, and C reprinted with permission from Elsevier Science.

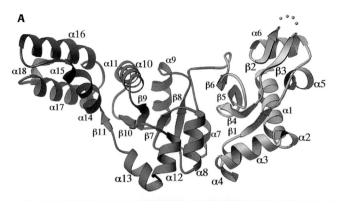

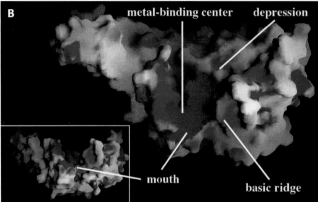

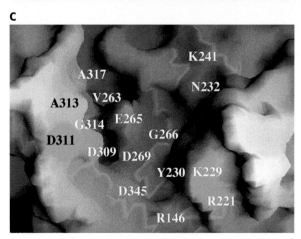

Figure 4.13. The 3-D structure of *E. coli* primase (DnaG). **A:** Ribbon diagram showing the three subdomains (red, blue, and yellow), with their beta strands and helices numbered. **B:** Surface potential view of the core domain (turned 90 degrees in the insert) indicating the mouth, the metal-binding center, and the depression. **C:** Active site with the metal-binding region showing the invariant (green) and highly conserved (yellow) amino acids. J. M. Berger, Science 287: 2482–6 (2000). Reprinted with permission from American Association for the Advancement of Science.

Figure 4.14. Model of interactions between the DnaG and DnaB helicase. The helicase is hexameric, but two parts have been cut away to reveal the interior central hole. In this model, the direction of RNA:DNA hybrid translocation and incoming ssDNA are opposite. Such opposition might restrict the length of the primer. J. M. Berger, Science 287: 2482–6 (2000). Reprinted with permission from American Association for the Advancement of Science.

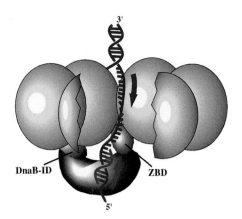

whole structure looks like a cashew, with the concave site characterized by a cleft. This cleft is dominated by a cluster of acidic metal-binding amino acids that form the mouth of the putative nucleic acid binding cleft (Figure 4.13C). It is possible that the single-stranded DNA is passing through the narrow mouth. Synthesis would occur at the metal-binding domain, and the DNA:RNA duplex would extrude into the depression. The structure of the core domain differs drastically from the basic palm metal-binding domain seen in many other nucleic acid polymerases. Perhaps this difference may account for the functional differences between primases and other polymerases, such as reduced processivity and lower fidelity.

After the primase has synthesized a primer, the primer remains tightly grasped by the primase and stabilized on the template by the SSB. The χ subunit of *E. coli* DNA polymerase III is known to bind to the same SSB, thus destabilizing the contacts between the primase and the SSB. This reaction leads to the assembly of the β clamp and to processive synthesis.

One possible model of interaction between DnaG and the hexameric DnaB helicase (structured as a ring; see next section) would place DnaG inward toward the center of the ring. The active mouth is positioned to accept the ssDNA as it exits from DnaB (Figure 4.14).

Processing of the Okazaki Fragments into a Continuous Strand

To create a continuous strand after Okazaki fragments have been synthesized, the RNA primers must be removed, the gapped region filled with deoxynucleotides, and the two adjacent 3′ and 5′ ends ligated. In prokaryotes, this procedure is possible by the action of DNA pol I that can remove the 5′ → 3′ RNA primers due to its 5′ → 3′ exonuclease activity and fill the gap using its polymerase activity. In eukaryotes, because DNA polymerase alpha and delta do not have a 5′ → 3′ exonuclease activity, removal of the RNA primers is mediated by a 5′ → 3′ exo/endo nuclease called FEN-1 or by Rnase H. After the completion of an Okazaki fragment its 5′-end is displaced by a helicase.

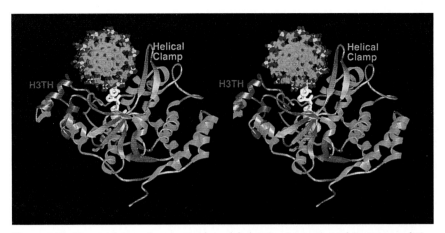

Figure 4.15. Stereo picture of a structural model showing interaction of *Pyrococcus furiosus* FEN-1 with the 5′ flap substrate of double-stranded DNA. The enzyme has a DNA-binding cleft, basically made of the helical clamp (green) that caps the active site and the H3TH domain that interacts with the phosphodiester backbone. The flap (white) passes over the active site (two metals, yellow spheres) and through the helical clamp. The C-terminus is labeled orange. J. A. Tainer, Cell 95: 135–46 (1998). Reprinted with permission from Elsevier Science.

This can generate a flap that is the substrate for FEN-1. Most likely FEN-1 is recruited to the site by PCNA because these two factors do interact. Figure 4.15 shows the structural interaction between FEN-1 and DNA.

FEN-1 also interacts with PCNA. This interaction is accommodated by the FEN-1 C-terminus. The interaction of the C-terminus domain seems to be similar to that of the p21 peptide because p21 competes with FEN-1 for binding to PCNA (see Figure 4.12). These interactions can be seen in Figure 4.16.

After the primer is removed, the gap is filled by a DNA polymerase, which leaves a nick that needs to be sealed by another enzyme, the DNA ligase. Ligase also seems to bind PCNA at the nick between two adjacent Okazaki fragments. DNA ligases form phosphodiester bridges between the 5′-phosphoryl and the 3′-hydroxyl groups of adjacent nucleotides at a nick in DNA. This reaction occurs in separate steps outlined in Figure 4.17. There are two classes of ligases, those that utilize ATP as a cofactor and those that utilize NAD^+. All eukaryotic DNA ligases are ATP dependent, while bacteria are NAD^+ dependent. In eukaryotes, there are four different ligases, but it seems that there is a strong preference for DNA ligase I. The three-dimensional structure of ATP-dependent T7 DNA ligase has revealed that the structure is characteristic of the nucleotidyl-transferase superfamily. The protein is composed of two domains with a deep cleft between them. A pocket that binds ATP is located in the N-terminus at the base of the cleft. DNA also seems to bind the cleft. More about the function of ligase can be learned by comparing this structure with the structure of the capping enzymes (see Chapter 8).

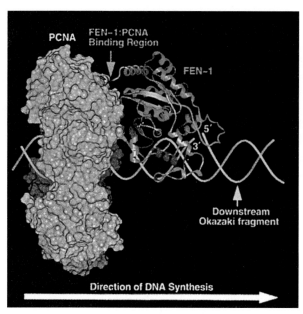

Figure 4.16. Model showing the interactions between PCNA, FEN-1, and DNA. FEN-1 is colored the same way as in Figure 4.15. Note the 5′ flap passing through the helical clamp and the interaction of the C-terminus with PCNA (arrow). J. A. Tainer, Cell 95: 135–46 (1998). Reprinted with permission from Elsevier Science.

A

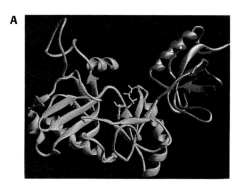

B $Enz - Lys - NH_2 + ATP \text{ (or } NAD^+) \longrightarrow Enz - Lys - AMP$

$$Enz - Lys - AMP + \overline{\quad OH \quad O \quad O5'\quad} \Longrightarrow \overline{\quad OH \quad O \quad O\quad}$$

Figure 4.17. A: 3-D structure of T7 DNA ligase. Domain 1 (N-terminal) is green and domain 2 (C-terminal) is red. ATP is shown in blue. D. B. Wigley, Cell 85: 607–15 (1996). **B:** Steps involved in the DNA ligation. First, the enzyme binds ATP with a lysine residue and forms an enzyme-AMP adduct (adenylation). Next, the adduct binds the 5′-phosphoryl group (transadelynation). Then, the adenyl group is displaced by the 3′-hydroxyl group, thus creating a phosphodiester linkage in the DNA strand. A reprinted with permission from Elsevier Science.

47

The Single-Stranded DNA-Binding Proteins

The single-stranded DNA-binding proteins are a very important component of the replication machinery. First, their major role seems to be protecting single-stranded DNA from degradation and aiding the helicase by disallowing reannealing of the complementary strands. Also, as mentioned previously, they are critical to the function and interaction of the primase. The three-dimensional structure of the single-stranded DNA-binding domain of human SSB, replication protein A (RPA) has been solved and has provided us with important clues of their interactions with DNA. RPA is a trimeric SSB and is highly conserved in eukaryotes. Its largest subunit, RPA70, binds to single-stranded DNA and mediates interactions with other cellular proteins. The DNA-binding domain is made up of two structurally homologous subdomains (A and B) oriented in tandem (Figure 4.18). In the structure (cocrystallized with an octadeoxycytosine), the DNA lies in a channel, which extends from one subdomain to the other (Figure 4.18). Three nucleotides are bound by each of the subdomains. The single-stranded DNA interacts with the RPA by stacking between bases and conserved amino acids and by hydrogen bonding. In fact, the stacking contacts between conserved phenylalanines are more prominent in subdomain A than B, and, in general, subdomain B seems to make fewer contacts with the single-stranded DNA.

Viewing the Replication Complex

The replication machinery is very complex. In the previous sections, the players of the replisome have been presented in 3-D structure individually. Also, interactions with the different players, when available at the structural level have also been visualized. Ideally, it would be very interesting if we could put all these components of the replisome together at 3-D. Something like that is impossible at this point. However, I believe it will be very useful to present general 3-D models and illustrations so that they can summarize all the events mentioned so far.

In Figure 4.19, we have all the major players during the synthesis of the leading and lagging strands (left panel). The reader needs no further explanation. All the major players were examined in detail in previous sections. Note, however, that the opening of the fork results in swivels of the strands (see section entitled "Topoisomerases"). In the right panel, the illustration pinpoints the switch between primase and polymerase activity. In stage A, a six-subunit helicase encircles the lagging strand and interacts to primase and the DNA pol III holoenzyme. Each core polymerase is bound to DNA by the β clamp, which is placed on the lagging strand by the γ complex. In stage B, the χ subunit of the clamp loader displaces the primase by interfering with the interaction of primase and SSBs. In stage C, the primase reconnects with the helicase and is ready to synthesize the next primer. Figure 4.20 is a 3-D model of this replication

A

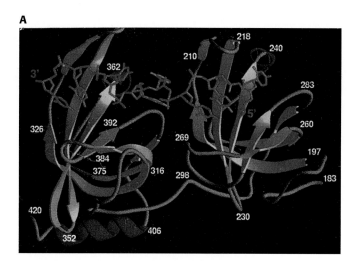

B

C

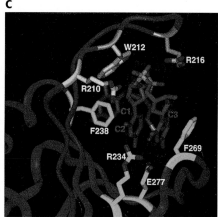

Figure 4.18. A: The RPA70/single-stranded DNA complex. Note the tandem orientation of the two subdomains with subdomain A on the right and subdomain B on the left. **B:** Same as in A but viewed down the axis of the single-stranded DNA-binding channel. **C:** Interactions of RPA subdomain A with the three deoxycytosines. Note the stacking between F238 and C1 and F269 with C3. Green amino acids interact with bases, and yellow amino acids interact with phosphates. L. Frappier, Nature 385: 176–81 (1997). Reprinted by permission from Nature, Macmillan Magazines Ltd.

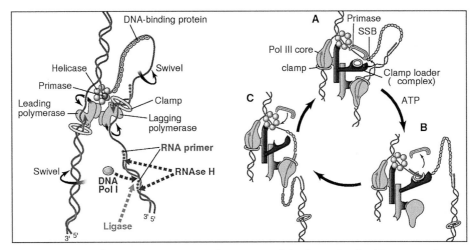

Figure 4.19. Illustration reconstructing the events and the factors involved in DNA replication. P. H. von Hippel, Science 287: 2436 (2000). Reprinted with permission from American Association for the Advancement of Science.

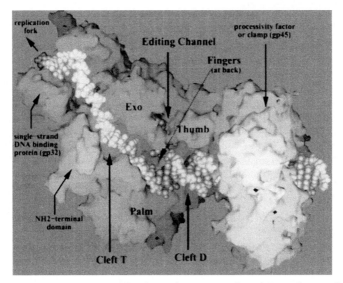

Figure 4.20. A 3-D model of the replication complex. Cleft T is the template binding, and cleft D is the duplex-product binding. T. A. Steitz, Cell 89: 1087–99 (1997). Reprinted with permission from Elsevier Science.

complex. This model is made with the coordinates of the 3-D structures of the RB69 DNA polymerase, the T4 SSB, gp32, and the RB69 clamp loader, gp45.

HELICASES

A very crucial process during DNA replication is the unwinding of the double helix at the replication fork. As DNA is replicated and the daughter strands approach the fork, the DNA must open and the two strands must dissociate to ensure continuation and progress of replication. The opening of the double-stranded helix is achieved by the action of helicases, which are bound to DNA near the fork and help open it up.

The general idea is that helicases destabilize the hydrogen bonds between the complementary strands in the double-stranded DNA in reactions coupled by the binding and hydrolysis of ATP. Obviously, many different helicases can also function in transcription and DNA repair. We assume that, in order for a helicase to unwind DNA, it must also translocate along the strands quite fast (maybe 500 to 1,000 bp/sec). Something like that suggests that helicases must move through conformational changes. All helicases appear to be dimers or hexamers, and thus have multiple DNA-binding sites.

Helicases can be divided in different groups depending on their structure. For example, the *E. coli* Rep, the *Bacillus stearothermophilus* PcrA, and the hepatitis C virus NS3 helicases are monomeric or dimeric and are characterized by seven conserved helicase motifs (I, Ia, II, III, IV, V, and VI). These motifs consist of residues located near the nucleotide-binding site and the interdomain cleft (see later). In other words, these motifs create the interface for nucleotide and single-stranded DNA (ssDNA) binding. These helicases are members of an ATPase superfamily. The hexameric helicases, such as the T7 helicase-primase or the *E.coli* DnaB, belong to another family that have only five motifs (I, Ia, II, III, and IV) of which only two (I and II) are similar to corresponding motifs found in the members of the ATPase superfamily. Next, the structures of different helicases will be presented to find clues about their function in DNA unwinding. In eukaryotes, unwinding of DNA is mediated by MCM proteins (see previous section entitled "Replication Origins in Eukaryotes").

The *E. coli* Rep Helicase

The solution of the 3-D structure of the *E. coli* Rep helicase provides important information into such function and for the process of DNA unwinding. The Rep helicase is a $3' \rightarrow 5'$ helicase. It has been shown biochemically to be a stable monomer in the absence of DNA, but dimers are induced upon DNA binding; therefore, its active form is a dimer. The Rep monomer consists of two domains (1 and 2) each divided into two subdomains (1A, 1B, 2A, 2B) (Figure 4.21). Domains 1A and 2A are composed of a large central parallel beta sheet flanked by alpha helices. The N-terminus is in subdomain 1A, and the C-terminus is in subdomain 2A. The structure of subdomains 1A and 2A shows similarities with the nucleotide-binding domain of the *E. coli* RecA protein (see later). The nucleotide (here is ADP)-binding site is located in 1A, and the DNA-binding site is in a cleft between subdomains 1B and 1A. The Rep monomer can exist as two configurations: an open and a closed one. The closed configuration is achieved by rotating domain 2B 130 degrees (Figures 4.21A and 4.21B). In Figures 4.21C and 4.21D, we can locate the conserved motifs, the

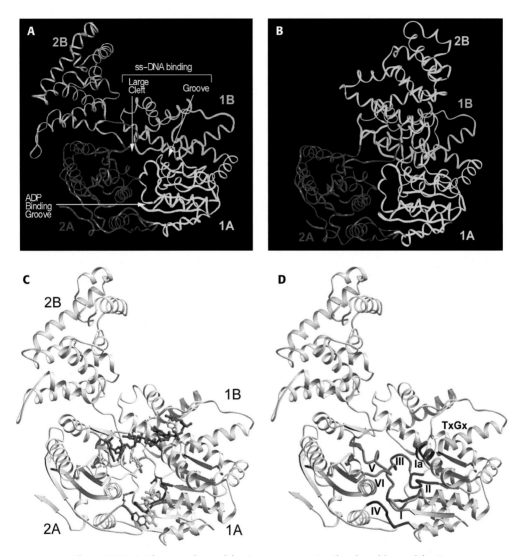

Figure 4.21. A: The open form of the Rep monomer. **B:** The closed form of the Rep monomer. **C:** Location of the ADP (green) and ssDNA (magenta) binding sites. Interacting residues with ADP are shown in orange and with ssDNA in yellow. **D:** Location of the helicase conserved motifs. Color code is I (light blue), Ia (magenta), II (dark blue), III (orange), IV (brown), V (purple), and VI (green). Light green indicates the TxGx sequence found in many helicases of the ATPase superfamily. G. Waksman, Cell 90: 635–47 (1997). Reprinted with permission from Elsevier Science.

binding of the nucleotide, and the ssDNA. Nucleotide binding involves motifs I and IV, while ssDNA binding involves motifs Ia, III, and V.

A functional Rep dimer can be generated by an open and a closed conformer. When the closed part of the dimer is rotated 130 degrees about hinge 1, the rotation results in a large motion of the ssDNA-binding site of the open conformer by 75 Å. This movement and the structural changes could account for the translocation of the helicase along the DNA, resulting in unwinding (Figure 4.22).

The *Bacillus stearothermophilus* PcrA Helicase

A different model has been proposed for the PcrA helicase from *Bacillus stearothermophilus*. When crystallized, this protein showed two different conformations with or without substrate DNA. Although the overall structure of the PcrA is similar to the closed structure observed for Rep helicase (with subdomains 1A and 2A similar to RecA), in the PcrA/DNA complex there were several different conformational changes (Figure 4.23). Therefore, it seems that for PcrA there are internal conformational changes that do not suggest the occurrence of dimers, and they do not seem to unwind DNA by a rolling mechanism as the one proposed for Rep. Rather, these changes suggest an inchworm mechanism whereby unwinding of DNA occurs (Figure 4.24).

The interaction of DNA with PcrA at the 3-D level reveals an impressive complementarity for the duplex binding and unwinding (Figure 4.25). The DNA duplex with a conformation close to the B-form DNA is bound to a groove on the surface of domains 1B and 2B (from the top up to A30-T5 pair). Halfway down the duplex to the junction (interface of double- and single-stranded DNA; G28-C7 pair), we can see a more distorted conformation. At the junction, the base pairs have begun to separate, the 3′ tail twists away into the center of the helicase, and the 5′ end of the other strand is placed across the outer surface of domain 2B.

The Bacteriophage T7 Helicase Domain

Both Rep and PcrA helicases are members of the ATPase superfamily and are monomeric or dimeric as explained earlier. However, another type of helicases is hexameric. The T7 helicase-primase and the *E.coli* DnaB are examples of hexameric helicases. In T7, the helicase-primase protein is encoded by one gene and consists of two rings, one large ring corresponding to the helicase domains and one small ring corresponding to the primase domains. This configuration is different from the arrangement of *E. coli* DnaB where the hexameric helicase ring is associated with one primase molecule (see Figure 4.14). Both domains are made up of six molecules or subunits. Each subunit consists of a mononucleotide-binding fold that resembles the corresponding one in *E. coli* RecA. Also, this domain is structurally similar to the corresponding domain 1A or 2A in Rep and PcrA helicases. In Figure 4.26, we can see the structural similarities between a T7 helicase subunit and RecA, as well as the arrangement of the subunits in a hexameric ring.

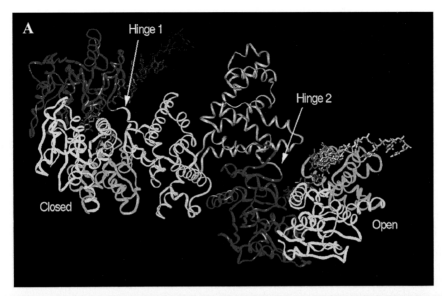

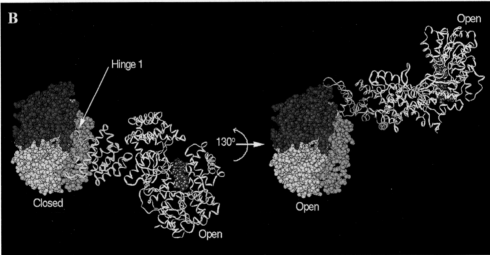

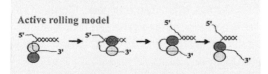

Figure 4.22. A: A possible subdomain 2B-mediated Rep dimer generated by an open and a closed monomer. The DNA is shown as single stranded and is colored purple. Opening the closed monomer would result in moving the ssDNA of the previously open monomer 75 Å. **B:** This activity could account for translocation and subsequent unwinding via a rolling mechanism as shown in the illustration below the 3-D models. G. Waksman, Cell 90: 635–47 (1997). Reprinted with permission from Elsevier Science.

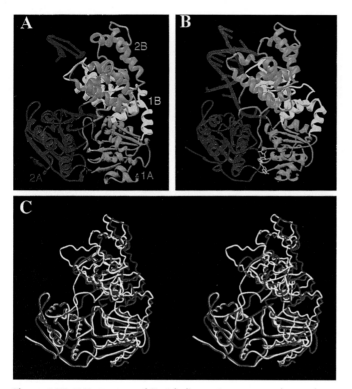

Figure 4.23. 3-D structure of PcrA helicase. A represents the product complex, and B denotes the helicase complexed with the substrate. The bound DNA is colored magenta. The difference in the two conformations can be seen in the stereo image in C. The product complex is red, and the substrate complex is white. Major changes can be seen in the interface between subdomains 1A, 2A, and 2B. D. B. Wigley, Cell 97: 75–84 (1999). Reprinted with permission from Elsevier Science.

Without 3-D structures of T7 helicase complexed with DNA, we do not have a clear picture of the mechanism of unwinding. However, based on the structure, it is possible that both the inchworm and the rolling mechanisms can be employed. In the inchworm mechanism, two adjacent RecA-like domains can interact with DNA in the same way that a pair of RecA-like domains interact in the PcrA monomer. However, the DNA-binding domain in T7 is located in a different region than the DNA-binding surface in PcrA. Alternatively, the similarity in sequence and location of the DNA-binding motif between T7 helicase and F1 ATPase could suggest an active rolling mechanism of unwinding.

The 3-D structure of a longer fragment of the T7 gene 4 protein (fragment 4D) provides some important information on the process of DNA translocation. The hexamer structure shows a deviation from the sixfold symmetry, which could correspond to an intermediate during catalysis. Also, the structure of a complex with nonhydrolyzable ATP analog showed that only four out of the six possible nucleotide-binding sites were occupied in the asymmetric

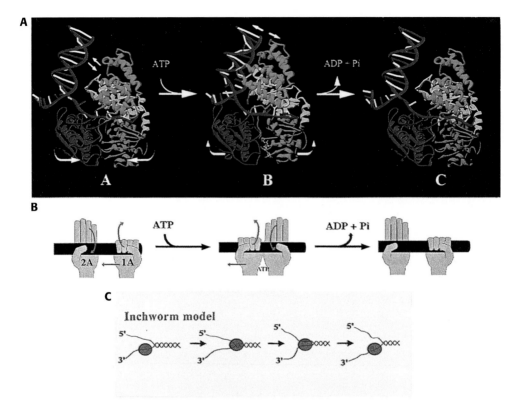

Figure 4.24. A: Models for PcrA helicase activity based on the structural changes observed upon DNA binding. In the first step, the protein is bound to the ssDNA tail and not the double helix. Upon ATP binding, there are conformational changes, and the DNA duplex binds subdomains 1B and 2B. At the same time, there is active unwinding at the junction. After ATP hydrolysis, the protein returns to its previous conformation. **B:** The changes in PcrA structure during the process. An open hand is a loose grip on the DNA and a closed hand is a tight one. **C:** The inchworm model for PcrA function. D. B. Wigley, Cell 97: 75–84 (1999). Reprinted with permission from Elsevier Science.

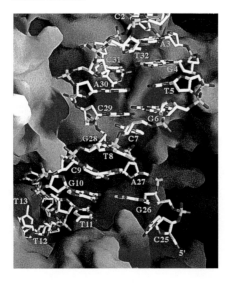

Figure 4.25. Interactions between PcrA helicase and DNA. The upper part shows the binding of the duplex, and the lower part shows the resulting paths or the single-stranded DNA after unwinding. The surface of the protein is colored red for negative electrostatic potential and blue for positive. D. B. Wigley, Cell 97: 75–84 (1999). Reprinted with permission from Elsevier Science.

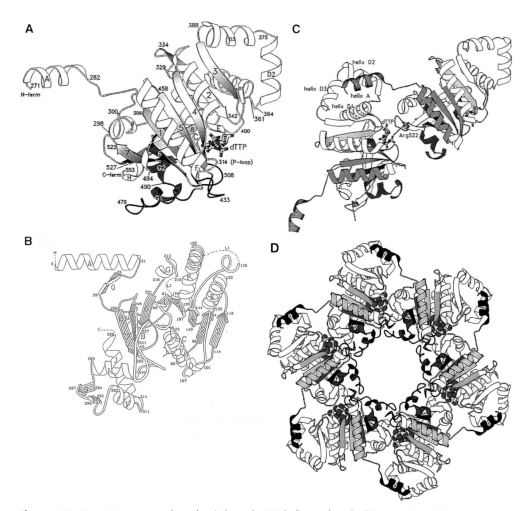

Figure 4.26. A: 3-D structure of a subunit from the T7 helicase domain (T7 gene 4 protein fragment 4E). Helices G1 and G2 (purple) comprise motif IV, which is part of the DNA-binding surface of the helicase domain. The nucleotide (here dTTP) is bound to the P loop, and motif II located at the end of helix E. T. Ellenberger, Cell 99: 167–77 (1999). **B:** 3-D structure of *E. coli* RecA to indicate the structural similarities. Note the similarity in the nucleotide-binding site. Here the nucleotide (ADP) is bound to helix C. DNA is bound by loops L1 and L2. Courtesy of T. A. Steitz, R. A. Welch Foundation (1992). **C:** Interactions between neighboring subunits in T7 helicase is mostly restricted between helices D1, D2, and D3 of one subunit and the N-terminal helix A of the other subunit. There are also some interactions between loops around the nucleotide-binding site (DNA-binding loops). The different conserved motifs have been colored as follows: I (red), Ia (yellow), II (green), III (blue), IV (purple; dark blue in D). T. Ellenberger, Cell 99: 167–77 (1999). **D:** View of the symmetric hexamer. T. Ellenberger, Cell 99: 167–77 (1999). A, C, and D reprinted with permission from Elsevier Science.

conformation of the hexamer. This suggests a "binding change" mechanism to explain unwinding of the duplex (Figure 4.27).

In this chapter and in other places throughout the book we have described the structure of NTP-binding proteins. In fact, many different proteins, often unrelated in function, contain an NTP-binding domain. These domains include proteins involved in such diverse duties as nitrogen fixation, proton-dependent ATPases, signal transduction, ribosome translocation, formation of Holliday junctions in DNA, and unwinding of DNA and RNA. These proteins hydrolyze ATP or GTP, and they have a common structural mechanism to achieve this. This mechanism depends on the presence or absence of the gamma phosphate, which is cleaved during hydrolysis. The members of this family of ATPases or GTPases contain a conserved GXXXGKT sequence at the NTP-binding site, and they orient ATP or GTP in an identical manner at the end of an alpha helix and loop containing this sequence. The gamma-phosphate interacts with a side chain of the backbone of an adjacent beta strand and loop, and this interaction is very conserved at the structural level (Figure 4.28).

TOPOISOMERASES

The active unwinding at the fork, catalyzed by helicases, can create real problems ahead of the fork. Imagine trying to pull two strings (such as in the DNA) to open them. The two strings ahead of the fork would become more compacted, supercoiled, and knotted. The best example is the spiral telephone cord, or any other cord, which with time does exactly that. Especially for circular DNA, the problem is even more severe. Because DNA is a helix, as a circular chromosome undergoes replication, the strands must rotate in order to separate. This means that a compensating winding occurs elsewhere in the circular DNA. The results of the stress would be that the circular DNA forms a supercoil (Figure 4.29). For the DNA duplex, however, this tendency is not good news because such topological distortion would result in premature termination of replication. To alleviate these problems, enzymes, called topoisomerases, are able to nick superwound DNA and, therefore, relax it, bringing it back to normal conditions, which allow continuation of unwinding and replication.

Two major types of topoisomerases can nick DNA. Type I topoisomerases are able to remove one supercoil (one double-strand break) and type II can remove two supercoils (two double-strand breaks).

Type I
in prokaryotes:
    Removes one negative supercoil per reaction cycle
    Requires no ATP
    Attaches to a 5′ phosphoryl group via a tyrosine residue
    Turns DNA one time around the nick, which removes phosphate
    Closes nick in the DNA by DNA ligase

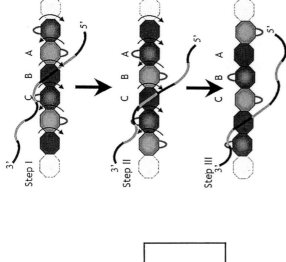

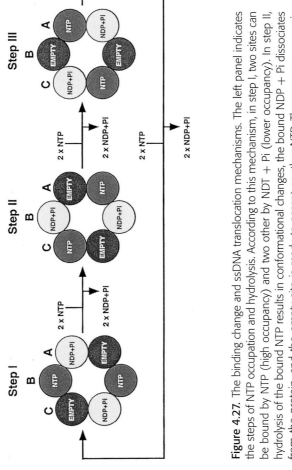

Figure 4.27. The binding change and ssDNA translocation mechanisms. The left panel indicates the steps of NTP occupation and hydrolysis. According to this mechanism, in step I, two sites can be bound by NTP (high occupancy) and two other by NDT + Pi (lower occupancy). In step II, hydrolysis of the bound NTP results in conformational changes, the bound NDP + Pi dissociates from the protein, and the empty site is ready to accept another NTP. The same sequence is repeated to progress to step III, but it takes place in different subunits. In this sense, ATP hydrolysis can be regarded as a ripple going around the ring. This sequence would not require rotation of the ring. The DNA would be translocated as it sequentially binds the occupied monomers (right panel). The DNA is represented as orange and black bands, each one indicating the step size for translocation. The DNA binding loops are red. The representation in the right panel is with the subunits laid out in two dimensions for clarity. D. B. Wigley, Cell 101: 589–600 (2000). Reprinted with permission from Elsevier Science.

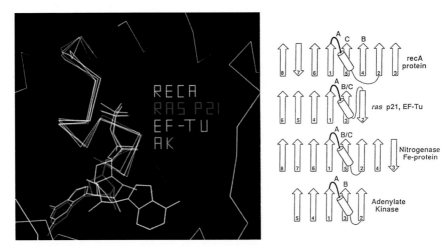

Figure 4.28. Comparison of four proteins that have the ability to bind nucleotide. In the left panel, the alpha carbon atoms of 12 residues located on the G/AXXXXGKT/S consensus sequence and the corresponding bound nucleotide from four proteins are superimposed. EF-Tu: elongation factor Tu; AK: adenylate kinase. Note the structural similarity in all proteins. Only the ring of the gamma-phosphate in the RecA complex differs. In the right panel, the beta sheet topologies of the enzymes are shown. The phosphate-binding loop is designated as A and follows the first beta strand. T. A. Steitz, R. A. Welch Foundation (1992). Courtesy of Dr. T. A. Steitz.

Type I
in eukaryotes:
    Can remove a negative or a positive supercoiling
    Attaches to 3′ phosphoryl group
Type II
in eukaryotes:
    Removes both negative and positive supercoils per cycle

Figure 4.29. The left panel illustrates a circular DNA molecule halfway in replication from a bi-directional origin (Ori). The parental strands are black, and the daughter strands are red. The illustration in the middle is a replication intermediate with three supercoils in the unreplicated region. The photograph on the right is such an intermediate observed by electron microscopy. An illustration of this intermediate is also shown beside the photograph. N. R. Cozzarelli, Cell 94: 819–27 (1998). Reprinted with permission from Elsevier Science.

Attaches to 5′ phosphoryl group via a tyrosine residue

Needs ATP

Is used for catenation and decatenation

Has topo III as bacterial homolog

Another type of topoisomerase II, found only in bacteria, is the gyrase, which actually introduces a negative coil. Therefore, the actions of topoisomerase and gyrase could control the rate of replication.

The 3-D Structure of Human Topo I

The discovery of the 3-D structure of human topoisomerase I in complex with DNA have provided insights into how a topoisomerase can relax DNA. The structure of topo I is shown in Figure 4.30. B is a rotation of 90 degrees about the vertical axis related to A. Topo I has three core subdomains (designated I, II, III), a C-terminal domain, and a linker domain, which is not presented in Figure 4.30. As can be seen in B, topo I has a central pore with a diameter of about 20 Å, which is a highly positively charged region for DNA binding. A model representing DNA binding to topo I at the configuration shown in Figure 4.30A is shown in Figures 4.30C and 4.30D without and with the linker domain, respectively.

Topo I, as expected, does not show any sequence specificity. However, a sequence from the ribosomal DNA of *Tetrahymena* is a high-affinity binding and cleavage site for eukaryotic topo I. The preceding models as well as those in Figure 4.31 are all made with the following sequence:

$$
\begin{array}{ccccccccccc}
 & -5 & -4 & -3 & -2 & -1 & +1 & +2 & +3 & +4 & +5 \\
5'\text{-} & G & A & C & T & T_\wedge & A & G & A & A & A\text{-}3' \\
\end{array}
$$
cleavage site

Topo I binds to this sequence at the 3′ phosphoryl group of the T (−1) as follows:

Figure 4.31 presents a close view of the interaction between topo I and DNA.

Mechanism of Cleavage and Relaxation of DNA

To study the interactions with 3′ phosphate at the cleavage site, data have been received from the 3-D solution of a wild-type and a mutant molecule, where wild-type Tyr-723 has been replaced with Phe. This transition renders the

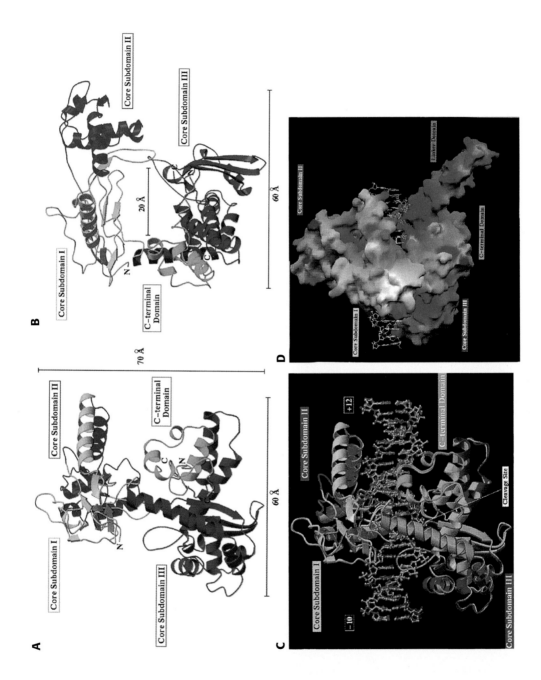

A

Core Subdomain I   Core Subdomain II
C-terminal Domain   Core Subdomain III
70 Å   60 Å
N

B

Core Subdomain I   Core Subdomain II
Core Subdomain III
C-terminal Domain
20 Å   60 Å
N   C

C

Core Subdomain II   +12
Core Subdomain I   C-terminal Domain
Cleavage Site
Core Subdomain III
– 10

D

Core Subdomain II   Linker Domain
Core Subdomain I   C-terminal Domain
Core Subdomain III

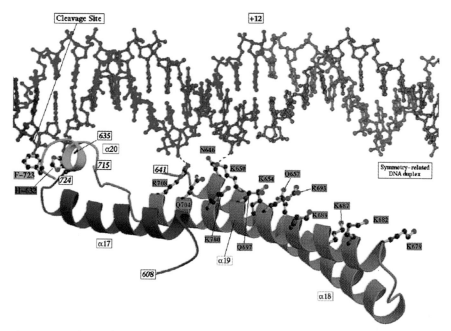

Figure 4.31. The catalytic site of topo I depicting helices 17, 18, 19, and 20. Helix 17 belongs to subdomain III, while helices 18 and 19 belong to the linker region. Helix 20 is part of the C-terminal domain and contains the catalytic Tyr723 (shown as Phe723), which forms a phosphodiester bond with the 3′ phosphate at the site of the cleavage of the scissile strand. W. G. S. Hol and J. Champoux, Science 279: 1504–13 and 1534–41 (1998). Reprinted with permission from American Association for the Advancement of Science.

enzyme inactive (observation of the structure before cleavage is then possible) and the binding with DNA noncovalent (Figure 4.32), while the binding with the normal Tyr-723 molecule is the covalent complex (observation after cleavage is possible, Figure 4.33).

The structural features presented in Figures 4.32 and 4.33 provide several suggestions and insights into the mechanisms of catalysis by eukaryotic topo I. Obviously, the tyrosine hydroxyl group is located close enough and is positioned for the nucleophilic attack and covalent attachment to the 3′ end of the nicked strand. The conserved Arg-488 and Arg-590 should stabilize the pentavalent coordination through hydrogen bonding to one of the nonbridging oxygen atoms of the scissile phosphate. The other nonbridging oxygen should stabilize through hydrogen bonding to the Nalpha2 atom of His-632. Most likely, His-632, which does not contact DNA in the noncovalent complex, contributes to the cleavage by donating a proton to the 5′ leaving group (Figure 4.34). Also the

Figure 4.30. The structure of human topoisomerase I viewed perpendicular to the pore (A) and rotated 90 degrees (B). **C:** A model of topo I with bound DNA. **D:** Same model with the linker domain. W. G. S. Hol and J. Champoux, Science 279: 1504–13 and 1534–41 (1998). Reprinted with permission from American Association for the Advancement of Science.

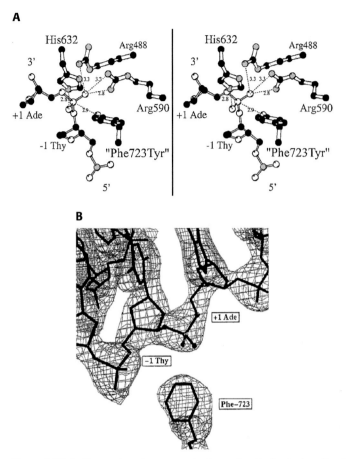

Figure 4.32. A: Stereo view of a noncovalent complex. Hydrogen bonding (dotted lines) between the catalytic residue (Phe-723-Tyr), His-632 (from subdomain III), Arg-488, and Arg-590 (from subdomain III) with the oxygen atoms of the phosphate group. **B:** Electron density maps near the active site of the noncovalent complex. Note that, in this complex, Phe-723 does not make contact and a normal phosphate bond exists between the 5′ Thy (−1) and the 3′ Ade (+1). W. G. S. Hol and J. Champoux, Science 279: 1504–13 and 1534–41 (1998). Reprinted with permission from American Association for the Advancement of Science.

enzyme contacts the phosphate groups only, and not much interactions is seen at the sequences downstream the cleavage site.

The structure of topo I and its interaction with DNA add to the debate about the possible mechanism of DNA relaxation. Earlier models suggested that topo I functions either by allowing the free 5′ end to rotate freely about the unbroken complementary strand or by imposing a tight fist on the DNA downstream of the cleavage site. Both of these models are not supported well by the structural data. First, the free rotation would need more extensive structural alterations. Second, because the interaction of the enzyme with DNA occurs mainly with the phosphate groups and not much interaction is seen downstream of the cleavage site, a tight grip on the DNA by the enzyme is not supported. Rather,

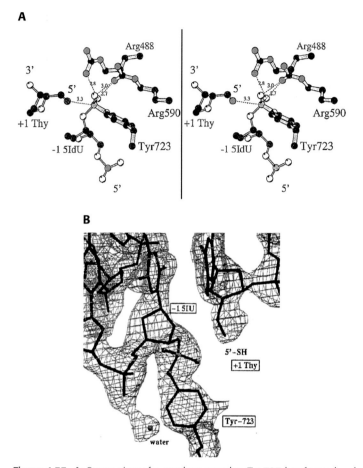

Figure 4.33. A: Stereo view of a covalent complex. Tyr-723 has formed a phosphodiester bond with the 3′ phosphate at the cleavage site. In this covalent complex, the duplex is made with a bridging phosphorothionate (from 5-iodo-deoxyuridine), a diester that can be cleaved by the enzyme, but whose fragments cannot rejoin. By doing this, the cleaved form of the complex can be received and observed. The cleavage is now between the 5′ 5IU (−1) and the 3′ Thy (+1). Note the phosphodiester bond of Tyr-723 with the 3′ end of the scissile strand. **B:** Electron density maps of the covalent complex. Note the presence of the free 5′-sulfhydryl (SH). Numbers indicate the distance in angstroms. W. G. S. Hol and J. Champoux, Science 279: 1504–13 and 1534–41 (1998). Reprinted with permission from American Association for the Advancement of Science.

it seems that DNA is relaxed by "controlled rotation" where small rocking movements are allowed and lead to relaxation (Figure 4.35).

The Structure of Yeast Topoisomerase II

As mentioned in the previous section, topoisomerase II can make two breaks in the DNA. Therefore, they can be used for catenation and decatenation, which entails the passing of one duplex through another. The breaking in each DNA strand is mediated via a transesterification using tyrosine

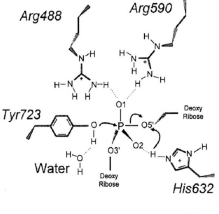

Figure 4.34. A possible intermediate stage in the cleavage by topo I. W. G. S. Hol and J. Champoux, Science 279: 1504–13 and 1534–41 (1998). Reprinted with permission from American Association for the Advancement of Science.

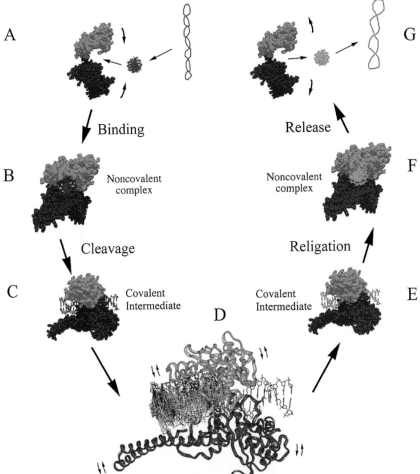

Figure 4.35. The controlled rotation mechanism for human topo I. Supercoiled DNA is red, and relaxed DNA is green. W. G. S. Hol and J. Champoux, Science 279: 1504–13 and 1534–41 (1998). Reprinted with permission from American Association for the Advancement of Science.

residues. The two breaks are four nucleotides apart. The eukaryotic enzymes are dimers, each monomer consisting of an N-terminal ATPase domain. The rest is separated in subfragments B' and A', which are homologous to gyrase B and gyrase A. Therefore, the prokaryotic (gyrases) enzymes are tetramers with two A and two B subunits. Let us now examine the structure of topoisomerase II encompassing the chain of the two subfragments, B' and A'. The B' subfragment consists of two alpha/beta domains. The A' subfragment has two parts. The N-proximal part of A' contains two domains and a connector. One domain not only has a fold similar to the CAP DNA-binding domain (see Chapter 5) and histone 5 but also contains the tyrosine that covalently attaches to the 5' end of the cleaved DNA. The second domain is an alpha/beta structure. The connector joins the two domains and the other part of A'. The C-proximal part is mostly alpha-helical, with three of them being long and prominent. Two of them create a coiled coil that creates the primary dimer contacts at the bottom of the molecule (Figure 4.36). Overall, the structure of a monomer looks like a crescent, and two of them make a heart-shaped dimer with a large hole (55 Å) in the middle (Figure 4.36B). The active-site tyrosine is projected from a loop and nested between the B' and A' domains. Each active site is accessed by a narrow tunnel, which begins at the B'2-B'1-A'2 intersection. The domains of the first part of A' creates a groove of 20 to 25 Å, which funnels down to the active site. This region has positive potential and most likely accommodates the DNA. In the model of Figure 4.36D, the 4-nucleotide 5' overhang extends from the groove to the tunnel, where the active site is.

This structure of topoisomerase II, which is also similar to gyrase, suggests a model for their function (Figure 4.37). First, the enzyme binds a duplex, called the G-segment. This binding leads to conformation changes, which result in interactions between the two top regions of the monomers as depicted in step 2. Upon binding ATP and another duplex (T-segment), the G-segment is split by the A' subfragments. At the same time, the ATPase domains dimerize, and the T-segment is transported through the DNA break into the central hole. Then, the G-segment is resealed, and the T-segment is released by opening the dimer at the A'-A' interface at the bottom of the dimer. Finally the A'-A' interface dimerizes again, and ATP is hydrolyzed.

TERMINATION OF REPLICATION

The replicative machine comes to a stop at specific sequences, which the so-called termination proteins use for binding. Termination sequences are present in both prokaryotes and eukaryotes, but the mechanism of the fork arrest in eukaryotes remains rather unknown. In prokaryotes, the termination sequences are found 180 degrees opposite to the replication origin. The terminator proteins cause replication arrest by stopping the progress of helicases. This function is called contrahelicase activity.

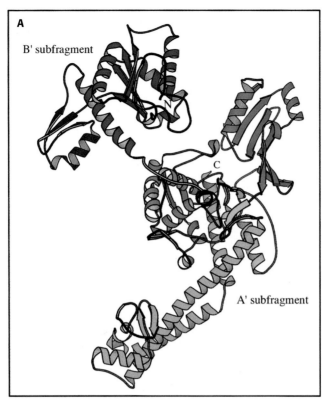

A

B' subfragment

N

C

A' subfragment

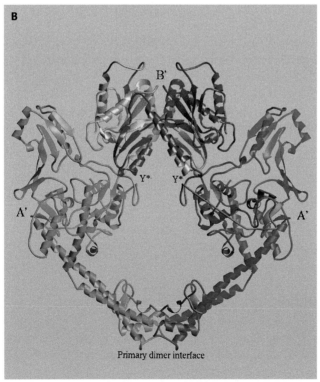

B

B'

Y*   Y*

A'   C   A'

Primary dimer interface

C

D

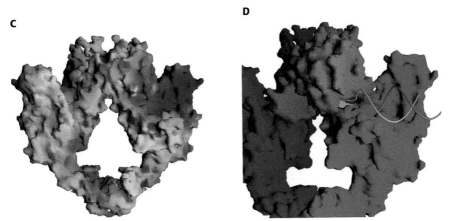

Figure 4.36. A: 3-D structure of one monomer of the yeast topoisomerase II (B', A'). The features that were described in the text are marked on the figure. B' is red, and N-proximal is purple (alpha/beta) and blue (CAP homology). The connector is light blue, and the C-proximal is green. **B:** The heart-shaped dimer of topoisomerase II. Y* is tyrosine. **C:** Surface representation of the heart-shaped dimer of topoisomerase II with the electrostatic potential. **D:** The intersection between B'2-B'1-A'2. The active site is and the tunnel where the 5' overhang is extended. The groove that accommodates DNA is also shown. J. M. Berger, Nature 379: 225–32 (1996). Reprinted by permission from Nature, Macmillan Magazines Ltd.

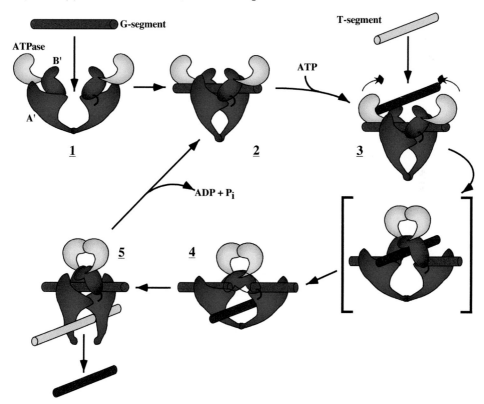

Figure 4.37. Model of the catalytic action of topoisomerase II as suggested by the 3-D structure and biochemical data. J. M. Berger, Nature 379: 225–32 (1996). Reprinted by permission from Nature, Macmillan Magazines Ltd.

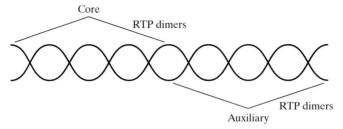

Figure 4.38. A terminus of replication.

In *Bacillus subtilis*, there are six termini. Each terminus consists of two overlapping binding sites: the core and the auxiliary sites (Figure 4.38). The core is bound by a dimer of replication termination protein (RTP). Then a second dimer binds to the auxiliary site.

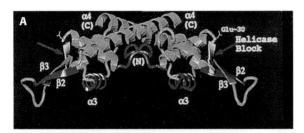

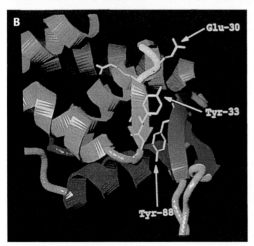

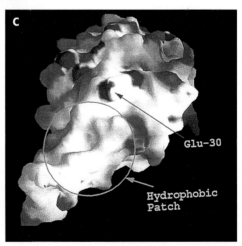

Figure 4.39. A: A ribbon model of the RTP dimer. The DNA-binding domain is shown in red, the dimer-dimer interaction part is blue, and the contrahelicase region is yellow. **B:** The contrahelicase region of RTP. **C:** The interaction site of helicase and contrahelicase. D. Bastia, Cell 87: 881–91 (1996). Reprinted with permission from Elsevier Science.

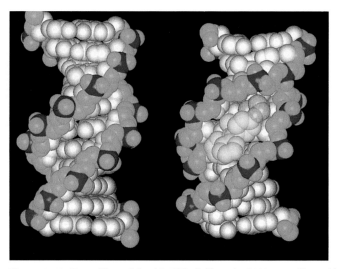

Figure 4.40. Spacefill models of B-DNA (left) and a DNA-tamoxifen adduct (right). Tamoxifen is yellow. Note the insertion into the minor groove, which results in enlarged and bent DNA. D. J. Patel, J. Mol. Biol. 302: 377–93 (2000). Reprinted with permission from Academic Press Ltd.

The 3-D structure of the contrahelicase region of RTP suggests a mechanism whereby the helicase-contrahelicase interaction results in termination (Figure 4.39A). In Figure 4.39B, we can see an enlarged view of the contrahelicase region. Two residues, Glu-30 and Tyr-33, are projected outwardly. The Tyr-88 is part of the dimer-dimer interaction strands and is projected inwardly. The arrangement of the amino acid residues in the contrahelicase region provides it with unique properties. In Figure 4.39C, we can see the electrostatic potential of this region as viewed by the approaching replication fork. The prominent region of contact with the fork and the docking site for the helicase is hydrophobic. Biochemical data have shown that this region interacts with the hinge region of helicases and possibly interferes with the rotational displacement of helicase on DNA. The final step in the termination is a decatenation process of the two intertwisted daughter strands by a topoisomerase. In *E. coli*, the Tus protein (RTP) interacts with DnaB and DNA for termination.

HOW DRUGS INDUCE MUTATIONS

We examined the process of DNA replication in this chapter and the architectural proteins that play roles in DNA replication and transcription in chapter 2. Many of these proteins interact with the minor groove and distort DNA. Also, we saw that DNA polymerase interacts with the minor groove of the base pair adjacent to the nascent base pair, which might account for the ability of DNA polymerase to scan for mistakes. Mutations due to drugs or chemicals might interfere with this process. Tamoxifen is a nonsteroidal antiestrogen that is an agent for treating breast cancer. However, it has also been linked to increased endometrial cancer, most likely because of its interaction with DNA. The 3-D structure of a DNA-tamoxifen adduct has shown that tamoxifen is inserted into the minor groove of DNA and results in enlarged and bent DNA (Figure 4.40). These alterations might affect the right geometry and the scanning of DNA polymerase and result in the inability of DNA polymerase to transfer the mismatched base to the exonuclease site, thus, enhancing the probability of mutations.

CHAPTER FIVE

Transcription in Prokaryotes

—— – – –

PRIMER Unwinding of DNA is not only restricted during replication but also occurs during transcription, even though the unwinding takes place in a different way. Transcription is a very important step, perhaps the most important step in regulation. Transcription determines which genes will be expressed and, therefore, proceed for decoding and protein synthesis. During the process of transcription, the $3' \to 5'$ noncoding DNA strand becomes the template for the synthesis of RNA. The RNA transcript, therefore, keeps the sequence order of the coding strand, $5' \to 3'$. The main enzyme here is RNA polymerase. Regulation at this level is very unique and markedly different in prokaryotes and eukaryotes. Therefore, for clarity, I describe transcription in prokaryotes first (this chapter) and continue with transcription in eukaryotes in Chapter 6.

The first part of this chapter examines the structure of the prokaryotic RNA polymerase and the mechanisms of transcription initiation and elongation as revealed by 3-D complexes of RNA polymerase and DNA. In the second part, I present some of the mechanisms involved in regulating gene transcription in prokaryotes. Having as a focus to present these events at the 3-D level, my aim is not to be redundant, and in doing so I have omitted information that can be found in microbiology and biochemistry books. Therefore, the basic structural motifs involved in transcriptional regulation are presented in a few regulators to pinpoint similarities and differences. Examples of the function of regulators from two bacteriophages is presented as revealed at the 3-D level. Termination of transcription makes up the final part of this chapter.

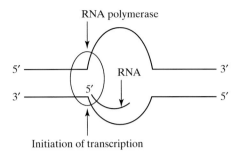

RNA polymerase

5′ ——————— 3′

3′ ——————— 5′

RNA

5′

Initiation of transcription

Figure 5.1. A general illustration of the process of transcription. At the initiation site, DNA opens up, and RNA polymerase initiates the 5′ → 3′ synthesis of RNA using the 3′ → 5′ strand (or noncoding strand) as template.

Transcription is the process during which RNA is synthesized. For this process, the double-stranded DNA must open up (at the transcription start sites) so that RNA can be synthesized using the 3′ → 5′ strand as template (Figure 5.1). In this sense, the process is similar to DNA polymerization, but the key enzyme is different. Also, during transcription and RNA synthesis, a ribonucleotide, rather than a deoxyribonucleotide as in DNA synthesis, is incorporated. The difference is that a nucleotide in the RNA has an OH group in the 2-position of the ribose. Transcription is a very complex event because it is intimately linked to gene regulation. The synthesis of certain RNAs is specific and demarcates areas of cell differentiation. As such, these events must be accurately controlled. This control is achieved by the actions of factors that interact with specific sequences of the gene that they help transcribe.

The key enzyme for the synthesis of RNA using DNA as a template is RNA polymerase. This enzyme has the ability to recognize certain DNA sequences by their initiation sites and to elongate the transcript. There are marked differences between the prokaryotic and eukaryotic transcription machinery. These differences include the number and the structure of RNA polymerases as well as the regulatory factors that synergize with the RNA polymerase. In this chapter, we will explore the structure and function of RNA polymerase as well as the mechanisms of regulation in prokaryotes. In Chapter 6, we will examine the eukaryotic transcriptional machine and regulation.

STRUCTURE AND FUNCTION OF PROKARYOTIC RNA POLYMERASE

In *E. coli*, there are two forms of RNA polymerase: the holoenzyme and the core. Both forms have four subunits in common (beta, beta′, two alphas; $\beta\beta'\alpha2$), but the holoenzyme contains the sigma factor in addition. The β and β' have molecular weights of 150 and 160 kDa, respectively. The α subunits are

40 kDa each, and the sigma factor is 70 kDa. The sigma factor directs the core to transcribe genes, allowing the RNA polymerase to bind tightly to the promoter region. It dissociates after initiation. The prokaryotic RNA polymerase is able to bind to the promoter regions of bacterial genes. These promoters contain sequences that are common in all promoters and are recognized by components of RNA polymerase. A typical bacterial promoter, called the core promoter, contains two conserved sequences: At position -10 the sequence is TATAAT, and at position -35 the sequence is TTGACA. Some strong promoters contain an extra element (called the UP element), which can be found upstream of the core promoter at position -40 to -60. The RNA polymerase extends from base -43 to base $+3$. The beta subunit is associated with base $+3$, and the sigma factor is near base -3. The C-terminus of the alpha subunit recognizes and binds the UP element.

There are many sigma factors, but the ones that are involved in the everyday growth are called primary sigma. When sequences from different sigma factors are aligned to find conserved sequences, four such regions can be identified. Region 1 (near the N-terminus) is found only in the primary sigma factors. Region 2 can be subdivided into four parts. Part 2.1 binds to the polymerase core, while part 2.4 recognizes the -10 box. Region 3 contains a helix-turn-helix DNA-binding domain, whose significance is unclear. Region 4 is sub-divided in two parts. Part 4.2 also contains a helix-turn-helix domain that binds to -35 box. All these interactions between alpha, beta, beta', and sigma are summarized in Figure 5.2. Sigma factor can also loosen the nonspecific interaction between the RNA polymerase and DNA and, consequently, enhance polymerase's specificity. Not all bacteria have this loosening activity from the sigma factor. In *B. subtilis*, for example, this activity is provided by factor delta.

The discovery of the three-dimensional structure of the prokaryotic RNA polymerase has shed light on its function and interactions with DNA and RNA transcript. The overall structure of the core and the holoenzyme differ. As can be seen in Figure 5.3, the overall structure looks like a hand with the fingers and the thumb creating a shape similar to the crab's claw. The holoenzyme is in the open form, and the core is closed. Most likely the open form of the holoenzyme allows it to grasp DNA to initiate transcription. After transcription is initiated, the sigma factor dissociates followed by this conformational change in core, which allows the enzyme to bind DNA very tightly so that it does not fall off during transcription and maintain high processivity. The closure creates a channel nearly 25 Å wide that fits double-stranded DNA perfectly. The images presented in Figure 5.3 are early electron microscope (EM) reconstructions and provide a good starting place for getting to know the structure. We now have much more detailed structures of RNA polymerase, and we will elaborate on these structures next.

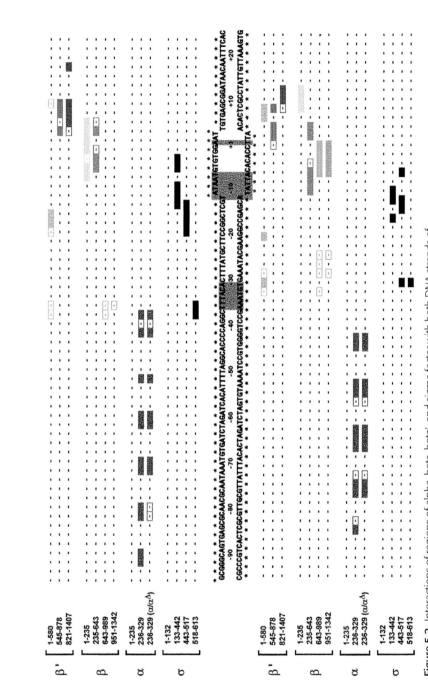

Figure 5.2. Interactions of regions of alpha, beta, beta', and sigma factor with both DNA strands of a bacterial promoter. The different color codes designate the corresponding region of each subunit. R. H. Ebright, Cell 101: 601–11 (2000). Reprinted with permission from Elsevier Science.

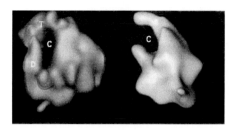

Figure 5.3. The *E. coli* core (left) and holoenzyme (right) as reconstructed from EM images at about 23-Å resolution. C, channel; T, thumb; D, bottom part of the closed channel. S. A. Darst, Cell 83: 365–73 (1995). Reprinted with permission from Elsevier Science.

THE 3-D STRUCTURE OF THE THERMUS AQUATICUS RNA POL

The 3-D structure of the *Thermus aquaticus* RNA pol was solved at a 3.3-Å resolution, which produced a clear image of the enzyme's 3-D structure. It does correspond well with the *E. coli* core crab claw structure. One arm of the claw is primarily the beta subunit, and the other is beta′. The width of the channel is 27 Å. To visualize the structure as clearly as possible, it is important to look at conserved domains by comparing sequences from many different prokaryotic RNA polymerases. Such a comparison is shown in Figure 5.4. The histogram reveals that there are some very conserved regions (A–H in beta′ and A–I in beta) as well as domains that denote the domain architecture in the structure (1–6 in beta′ and 1–8 in beta). From this comparison, it is obvious that beta and beta′ are also quite different, even though there is a certain similarity in sequence and structure.

Several rotated images of the 3-D structure are shown in Figure 5.5. Certain gross features can be realized immediately. The alpha domains have no access to the internal channel created by the beta (thumb) and beta′ (fingers) domain. Therefore, the alpha subunits should not have a role in the catalysis, but they are rather important in the beta, beta′ assembly. The omega subunit contacts the beta′ subunit only. It virtually wraps around the C-terminus tail of beta′. Maybe the omega subunit has a chaperonin role in the final stages of assembly. Regions of beta contact the alphaI N-terminus domain (NTD), and beta′ contacts the alphaII NTD. The C-terminus of beta (region betaI, domain 8 and part of 7 in Figure 5.4) makes extensive contacts with beta′. Beta domain 6 (region F-H) forms a flexible flap. Beta′ is virtually circular and its region E interacts with the beta subunit (Figure 5.6).

Let us now look at the catalytic site more closely. In Figures 5.5 and 5.6, we can observe beta′F beginning where beta′E ends. It forms a helical segment and loop that transverse across the middle of the main channel and end anchored in the main body of beta′. The active Mg^{2+} center is located at the base of the main channel directly across from the beta′F helix. The beta′F helix and the beta′G, which forms a loop, extend into the main channel, forming a wall-like structure that separates the main channel into two parts. The secondary channel is 10 to 12 Å, but it is not likely to hold the single-stranded DNA after melting. Between beta′F and beta′G there is a Zn^{2+}-binding domain. The active site (Figure 5.7) is the sequence NADFDGD belonging to beta′D. This sequence is absolutely

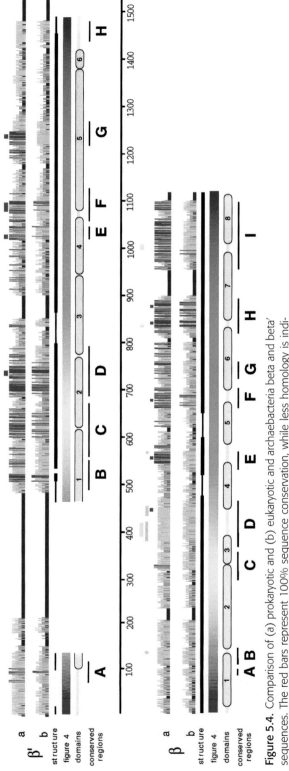

Figure 5.4. Comparison of (a) prokaryotic and (b) eukaryotic and archaebacteria beta and beta' sequences. The red bars represent 100% sequence conservation, while less homology is indicated by yellow, green and blue color bars, with blue bars indicating the least homology. The most conserved regions are indicated with letters A–I, and the domains in the structure are labeled with numbers 1–8. Note that the color bar is also used in Figure 5.6 to distinguish the different domains from N-terminus (red) to C-terminus (blue). S. A. Darst, Cell 98: 811–24 (1999). Reprinted with permission from Elsevier Science.

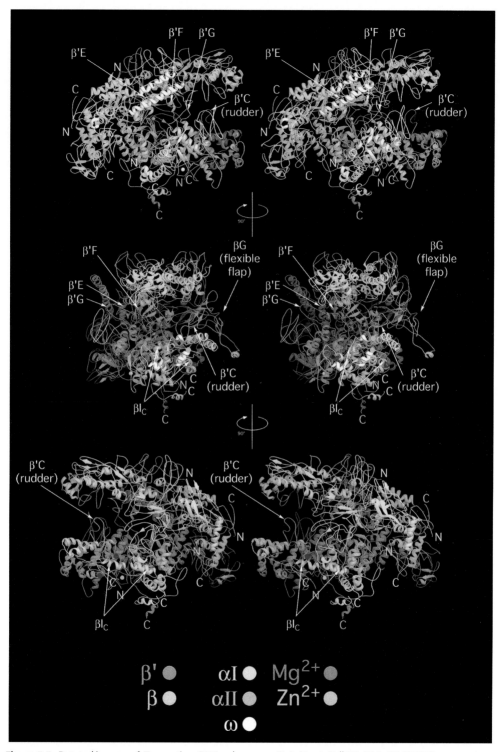

Figure 5.5. Rotated images of *T. aquaticus* RNA polymerase. S. A. Darst, Cell 98: 811–24 (1999). Reprinted with permission from Elsevier Science.

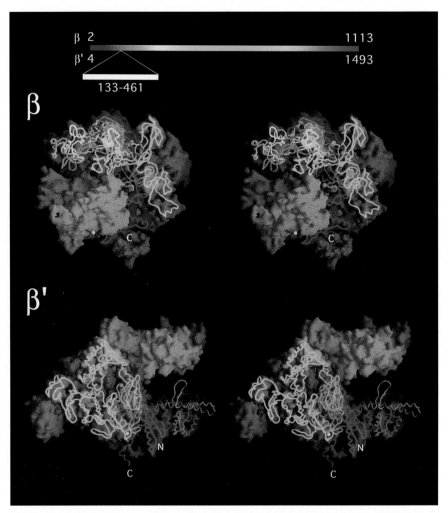

Figure 5.6. Stereo views of the beta and beta′ subunits following the color code of Figure 5.4. S. A. Darst, Cell 98: 811–24 (1999). Reprinted with permission from Elsevier Science.

conserved with the three Ds chelating the Mg^{2+} ion. The catalytic site has two sites for nucleotide triphosphate (NTP) binding. The i site, which becomes the 5′ end of the RNA transcript, and the i + 1, which is the elongation site extending the i site to the 3′ direction. Three residues from the beta subunit – Lys-838, His-999, and Lys-1004 – have been identified in close proximity (within a few angstroms) to the initiating NTP. The Rif site contains amino acids probably involved in binding DNA during initiation, but not during elongation.

Cross-linking mapping experiments using initiating NTP have shown interactions between the polymerase, DNA, and the RNA transcript. These interactions are depicted in Figure 5.8A. These images are open booklike sections through the inside of the channel. The left and the right images are the two

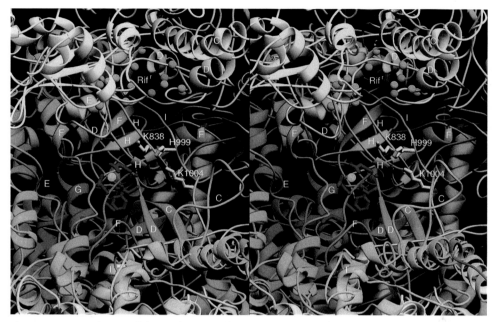

Figure 5.7. Stereo image of the catalytic site of *T. aquaticus* RNA polymerase. Beta subunit is colored cyan, beta' is rose, and al is yellow. The side chains of the conserved NADFDGD sequence are shown in red, and those of K838, H999, and K1004 appear yellow. The Mg^{2+} is shown as a pink sphere, while the magenta spheres depict the alpha carbons of amino acids in the Rif site. S. A. Darst, Cell 98: 811–24 (1999). Reprinted with permission from Elsevier Science.

"pages" of the book, and the top and the bottom images are the pages of an open writing pad. The top images are primarily from the beta subunit, and the bottom images from the beta' subunits. The parts of the proteins that have been sliced away are colored gray. The alpha-phosphate of the NTP in the i site interacts with residues colored yellow on the bottom. The gamma-phosphate of the NTP in the i site has been mapped to interact with a peptide on the upper face of the channel (yellow at top right), which coincides with the Rif site (magenta at top right). The template DNA strand interacts with residues colored green in the top and bottom right images. The upstream −10 base position of the RNA transcript interacts with the bottom of the channel near the "rudder" (blue residues on the bottom right). Based on all this, the orientation of DNA is shown by the white arrows. The DNA enters with its downstream and exits with its upstream portion. The downstream and upstream double helical DNA form an angle greater than 90 degrees. Figure 5.8B summarizes the structure of the transcriptional complex. The DNA template strand is denoted as a blue line, the nontemplate strand is shown as a cyan line, and the RNA transcript is a red line. From these studies, it seems that the main channel is occupied by 9 bp of dsDNA, about 9 bp of the RNA-DNA hybrid, and the nontemplate DNA strand (with unknown location). Room for the entry of NTP in the active

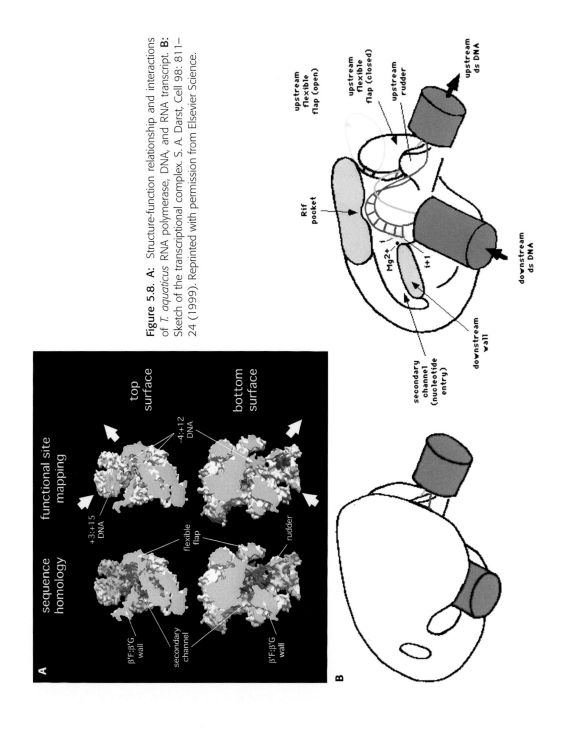

Figure 5.8. A: Structure-function relationship and interactions of *T. aquaticus* RNA polymerase, DNA, and RNA transcript. **B:** Sketch of the transcriptional complex. S. A. Darst, Cell 98: 811–24 (1999). Reprinted with permission from Elsevier Science.

Figure 5.9. This structure is at the same orientation as that in Figure 5.5, bottom panel. Aqua, beta subunit; rose, beta' subunit; gray, omega subunit; bright pink, Rif pocket; light purple, beta domain 3 (sigma contact b); yellow, rudder; green, beta' coiled coil (sigma contact a); dark purple, flap (sigma contact c). S. A. Darst, Cell 98: 811–24 (1999). Reprinted with permission from Elsevier Science.

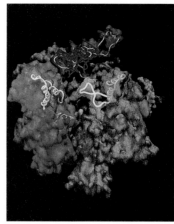

center cannot be accommodated. Therefore, it seems that the secondary channel allows access of the NTP substrates to the active center.

With the structural features of RNA polymerase clearly in mind, let us examine now the transitions in the transcription cycle. The holoenzyme first binds 40 to 60 bp of duplex DNA. After a series of structural modifications, the sigma factor dissociates, and the DNA template is placed in the active-site channel. From biochemical studies, certain sites of interactions of the sigma factor and the RNA polymerase have been characterized. One site (c in Figures 5.9 and 5.10)

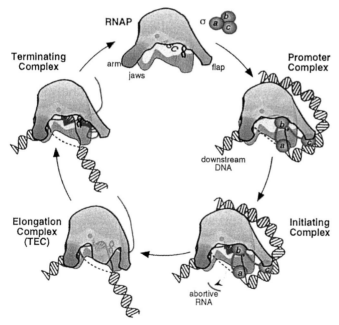

Figure 5.10. The sequence of events in prokaryotic transcription. R. Landick, Cell 98: 687–90 (1999). Reprinted with permission from Elsevier Science.

interacts with the tip of the flexible flap; sites a and b interact with the beta′ coiled coil that projects from the rudder on the lower surface of the active-site channel. These interactions most likely are responsible for closing the channel. The initiating complex results in RNA synthesis of only a few nucleotides, which constitute the abortive RNA. This RNA transcript is made while the sigma factor is still contacting the core. When it reaches nearly 8 to 9 nucleotides, the growing RNA chain encounters the sigma factor, which is bound to beta domain 3 and the beta′ coiled-coil. To extend the chain, these contacts must be lost; therefore, the sigma factor dissociates. After this stage, the channel closes, forming the elongation complex. At this point, the DNA is positioned well inside the active site, and the flap closes and creates the RNA exit tunnel that allows the RNA transcript to pass underneath it. This elongation complex is highly processive and, in bacteria, allows uninterrupted transcription for 10,000 bases without dissociation. In prokaryotes, transcription is terminated when a sequence of GC followed by a stretch of Us creates a termination hairpin. Most likely passing the termination hairpin under the flap destabilizes the elongation complex and results in opening of the RNA exit tunnel and the active-site channel. The structural features involved in these interactions are shown in Figure 5.9.

Figure 5.10 depicts the sequence of events from the binding of the sigma factor to the core to the termination. The core is gray, and the sigma factor is purple; the three sites of interaction with the core are denoted a, b, and c. The Mg^{2+} is a green sphere at the active site, and the channel through the enzyme is indicated by a lightened area. After sigma factor binding, the holoenzyme binds the promoter complex. DNA seems to wrap the enzyme, and the nontemplate strand is shown as a blue dotted line. The initiating complex results in synthesis of a short RNA that is aborted. Subsequently, the enzyme closes and forms a stable elongation complex of transcription while the DNA is positioned in the channel with a bend of 90 degrees. The RNA strand (red) passes under the flap whose closure has created the RNA exit tunnel. Formation of the termination hairpin wedged under the flap results in opening and termination.

In Figure 5.11, we can see the transcription elongation complex in 3-D for further clarification of the structures involved in the catalytic site and the path of the transcript. In Figure 5.11A, the template DNA strand is red, and the nontemplate strand is yellow. The RNA transcript is gold. Interactions around the DNA/RNA hybrid can be seen in a magnified version in Figure 5.11B. The positioning of the rudder suggests that it is involved in separating the exiting RNA from the DNA template strand. Indeed, cross-link studies have shown that RNAs longer than 9 nucleotides do not interact with the rudder (the DNA/RNA hybrid is 8 to 9 nucleotides long). The basic amino acids R598 and R601 are nearly invariant and may be involved in such function. Beta′F helix and betaD loops I and II (green in Figure 5.11A) might be involved in the flexibility of RNAP claws, which results in a close complementary fit between DNA, the DNA/RNA hybrid, and the main channel.

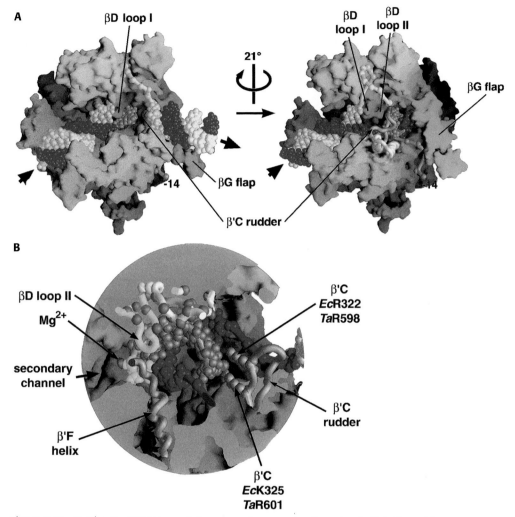

Figure 5.11. A: The *Taq* RNAP transcription elongation complex (two views). Note the rudder near the hybrid and the RNA transcript (gold) extruded via a channel formed by the flap. **B:** Interactions of RNAP parts with the hybrid. The Rif (cyan, backbone worms; green, beta loop II) extends over the top of the RNA transcript. The corresponding residues of the rudder from *E. coli* RNAP (Ec) are also shown. S. A. Darst, Science 289: 619–25 (2000). Reprinted with permission from American Association for the Advancement of Science.

Another interesting property of RNA polymerase is the fact that it pauses at certain sites along the DNA and halts transcription when it encounters the terminators in DNA sequences. It seems that RNA polymerases form slow or fast elongation complexes. In a slow complex, when a termination sequence is encountered, it pauses allowing an RNA hairpin formation, which results in the release of mRNA and termination. Faster complexes bypass these problems with the help of antiterminator proteins that stabilize the DNA-RNA hybrid. In other words, individual RNA polymerase molecules possess different intrinsic transcription rates and the ability to pause and stop. Also, it seems that single molecules can switch spontaneously to different rates. This could be the basis for another level of transcriptional control and regulation.

REGULATION IN PROKARYOTES

Bacteria contain only a few thousand genes. Some of these genes are active all the time; thus, elaborate regulation is not necessary. However, prokaryotes do regulate the expression of some genes that are turned off. These genes are mostly involved in metabolic pathways such as the synthesis of sugars and amino acids, which are necessary for survival. The bacterial cell has invented a very interesting and efficient way to regulate such genes. Genes that encode enzymes used in one metabolic pathway are grouped together and are regulated in a coordinated fashion. Such groups of genes are called operons. The organization of a typical operon and its regulation is depicted in Figure 5.12. This operon is made up of three genes (A, B, and C), which encode for enzymes needed for metabolizing a particular substance. In front of the three genes, a bacterial promoter that is necessary for RNA polymerase binding can be found. However, in the promoter, there is an element called operator (OR), which can be bound specifically by a repressor, R, which is encoded by a gene found upstream the promoter. The repressor gene can be a part of the operon, or it may not be. When the repressor is bound on the operator, RNA polymerase cannot be loaded onto the promoter, and there is no synthesis of the operon's mRNAs. The operon at this state is inactive. When the inducer, however, is present (which heralds the need for metabolism and the enzymes), it can bind the repressor and alter its 3-D structure. The altered repressor cannot any longer bind to the operator, and the promoter is now free for the RNA polymerase to bind and transcribe the operon. In the next pages, we will examine the structural basis of this remarkable type of regulation using a few examples. Then we will examine a different mode of operon regulation that depends more on termination hairpins (attenuator) than on repressors.

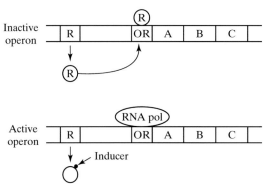

Figure 5.12. The basic features of a prokaryotic operon. A, B, and C are genes of the operon, R is the repressor, and OR is the operator site.

The Helix-Turn-Helix Motif in Prokaryotic Gene Regulation

A general feature in prokaryotic gene regulation is that many regulatory proteins (repressors, activators) bind the target DNA sequence with a very characteristic structural motif, the helix-turn-helix (HTH) motif. Despite the lack of sequence identity among the different HTH motifs, the fold is remarkably similar, even when it is compared to its eukaryotic counterpart (see Chapter 6). The second helix is the DNA recognition helix, which interacts with nucleotide bases at the major groove via hydrogen bonds and hydrophobic contacts. However, not all HTH motifs interact with the major groove in a similar manner. For some of them, such as the HTH motifs from the Lac repressor and the Trp repressor, the recognition helix is pointed in the major groove. For others, such as the HTH motifs of phage Cro and repressors, the recognition helix is embedded in the major groove. Interactions may also distort DNA. Also, some repressors interact with the DNA at the minor groove and distort DNA as well. In the next sections, we will examine some forms of regulation and present specific examples. These examples will clarify the basic ways of gene regulation in prokaryotes and the interaction of the regulatory proteins with DNA.

The Lac Operon

The Lac operon contains three genes (lacZ, lacY, lacA) that are responsible for the metabolism of lactose (see Figure 5.12). These genes code for the necessary enzymes to metabolize lactose to galactose and glucose. The three enzymes are beta-galactosidase (lacZ), permease (lacY), and transacetylase (lacA). In the absence of lactose, the repressor binds the operator very tightly, and the genes for lactose metabolism are not transcribed. When lactose is present, an inducer binds to the repressor and changes its shape. The inducer is an alternate form of lactose called allolactose. Allolactose is produced by beta-galactosidase, which after cleaving lactose can also rearrange and form allolactose. The allolactose-bound repressor cannot any longer bind the operator and frees the promoter making it accessible to RNA polymerase, which can now transcribe the genes that are needed for lactose metabolism. In fact, RNA polymerase does bind the repressor-occupied operator; therefore, it seems that the repressor inhibits the processivity of RNA polymerase. The operon is not 100% closed when the repressor is bound to the operator. Some leakage is possible so that the operon functions at some basal level that allows the presence of the enzymes and the production of allolactose.

The Lac repressor is encoded by the lacI gene, which is part of the Lac operon. The repressor protein is a monomer of 360 amino acids, but the functional complex is a tetramer. The repressor binds DNA via a helix-turn-helix motif, which points the recognition helix to the major groove. However, another helix, the so-called hinge helix, interacts with the minor groove. Such interaction bends DNA considerably (Figure 5.13).

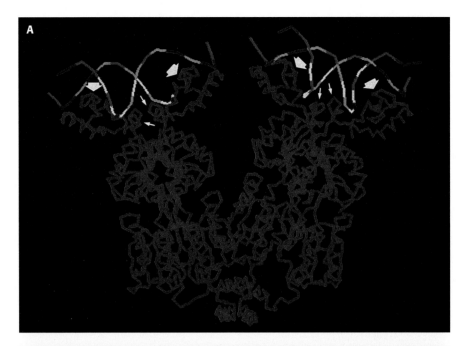

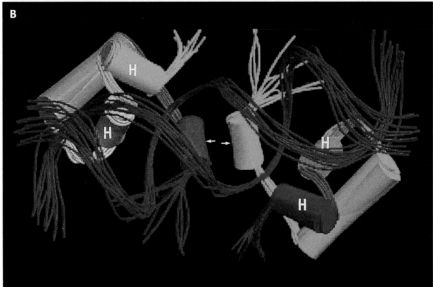

Figure 5.13. A: A tetramer of the *E. coli* Lac repressor bound to DNA. Note the HTH domain (thick arrows) interacting with the major groove and the hinge helix (thin arrow) interacting and distorting the minor groove. M. Lewis et al., NBD file PDR026, Science 271: 1247–54 (1996). **B:** A dimer of the Lac repressor DNA-binding domain bound to DNA via the helix-turn-helix motif (the helices are marked with H). The hinge helices are indicated by arrows. This is an NMR-determined structure; consequently, we see many different superimposed structures. C. A. Spronk et al., PDB file 1CJG, Structure 7: 1483 (1999).

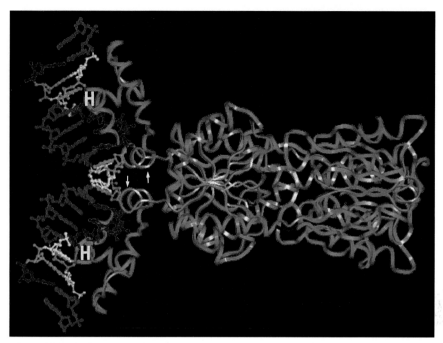

Figure 5.14. The interaction between an *E. coli* PurR dimer and its operator. Note the characteristic bending of the DNA due to the interaction of the hinge helix (arrows) from the minor groove. The recognition helix of the HTH motif is indicated by the letter H. M. A. Schumacher et al., NDB file PDR020, Science 266: 763 (1994).

The Pur Repressor

The Pur repressor controls the operon responsible for de novo synthesis of purine nucleotides. This complex pathway results in the synthesis of inosisate, whose purine base is called hypoxanthine and which can also bind the repressor. PurR belongs to the LacI family (as the Lac repressor) and binds the DNA with an HTH motif at the major groove and with its hinge helix at the minor groove. During interaction of an *E. coli*, PurR with its operators, leucine side chains interdigitate between central CpG bases and leads to a 45- to 50-degree kink. The resulting broadening of the minor groove increases the accessibility of six central base pairs for interactions with PurR (Figure 5.14). The reader should at this point recall other proteins that contact DNA from the minor groove and result in the bending of DNA, such as Sac7d, HMG (Chapter 2), and TBP (Chapter 6).

The Trp Repressor

Another example of a repressor is the Trp repressor, which controls genes necessary for synthesis of tryptophan. This repressor also has an HTH motif. The binding of this repressor to its operator is achieved by a conformational change after tryptophan binding. In this case tryptophan is considered to be

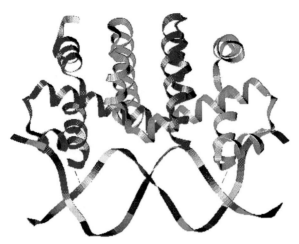

Figure 5.15. The interaction of *E. coli* Trp repressor with its operator. Note that each interacting helix is pointed in the major groove. Without tryptophan binding (inactive repressor), these helices are pointed more inward (broken lines). Z. Otwinowski et al., NDB file PRD009, Nature 335: 321–9 (1988).

a corepressor. Tryptophan binding alters the arrangement of the HTH motif by bringing the two interacting helices (from the dimers) within a distance of 34 Å, so that they can interact with two major grooves. The helices do not fit in the major groove as in the case of the 434 repressor, but they are pointed at the major grooves. As a result, only one amino acid interacts with a base; other interactions, direct or indirect (involving water molecules), are with the backbone phosphates (Figure 5.15).

Catabolite Activator Protein

Catabolite activator protein (CAP) assists RNA polymerase to bind more efficiently to certain *E. coli* promoters. CAP controls operons involved in the breakdown of sugars, such as the Lac operon. When the level of lactose breakdown is low, the concentration of cAMP (cyclic 3′, 5′ adenosine monophosphate) increases. This activates the CAP that binds to its specific operator and increases the rate of transcription in the adjacent operons. CAP is a nonspecific DNA-binding protein when not bound to cAMP; however, upon cAMP binding, CAP binds strongly to a specific operator. CAP is made up of two identical polypeptides of 209 amino acids, with each chain folded into two domains that have separate functions. The N-terminus domain binds cAMP and the C-terminus contains the helix-turn-helix motif, which binds DNA. The binding of the two chains induces a bend of nearly 90 degrees in the DNA (Figure 5.16).

REGULATION IN PHAGES

Phages, which infect bacteria, can regulate two major pathways. When a phage infects a bacterium, its DNA is incorporated in the host chromosome

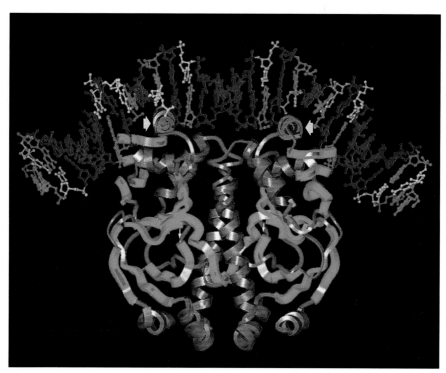

Figure 5.16. The 3-D structure of *E. coli* CAP bound to DNA. Note the two identical monomers, the cAMP binding (thin arrows), the HTH motif (thick arrows), and the bending of DNA. Alpha helices are purple, beta strands are light blue, and loops are dark blue. S. C. Schultz et al., NDB file PRD006, Science 253: 1001–7 (1991).

and can be replicated, but its genes cannot be expressed and stay dormant. The bacterium is then in the lysogenic state. However, by ultraviolet (UV) irradiation, *E. coli* can be stimulated to produce phage particles that result in lysis of the cells. In different bacteriophages such as the lambda, the 434, or the P22, a region of DNA is responsible for controlling the switch from the lysogenic to the lytic pathway. This region comprises two structural genes, Cro and the repressor that acts on six operator sites (three on the right – OR1, OR2, and OR3 – and three on the left – OL1, OL2, and OL3). The three operator sites have different preferences for the Cro or the repressor. The repressor protein binds to OR1 and OR2 (or OL1 and OL2), which is part of the Cro promoter, and such binding does not allow RNA polymerase to transcribe Cro. Under these conditions the bacterium is lysogenic. Cro protein can bind OR3 (or OL3), which is part of the repressor promoter. This binding results in inhibition of repressor transcription allowing the Cro to be made and lysis to ensue. This competition between two proteins and DNA sites is one of the best examples of how small changes in the different OR sites can account for differential affinity of protein binding to DNA and differential regulation. Let us dissect these molecular interactions and visualize such a regulation using the 434 and the lambda phage. The region of DNA that contains all the control elements (we will concentrate with the right, OR) has the following arrangement:

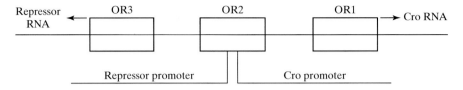

The 434 Phage

The OR1, OR2, or OR3 sites are very similar in sequence. There is only a small difference in OR3.

```
                  L        |        R
          1 2 3 4 5 6 7  | 7' 6' 5' 4' 3' 2' 1'
    OR1   A C A A G A A  | A G T T T G T
          T G T T C T T  | T C A A A C A
          1' 2' 3' 4' 5' 6' 7'  | 7 6 5 4 3 2 1


    OR2   A C A A G A T A C A T T G T
          T G T T C T A T G T A A C A


    OR3   A C A A G A A A A A C T G T
          T G T T C T T T T T G A C A
```

As we can see in this comparison of the sequences, the different OR sites are, in fact, palindromes. In other words, the right half-site is identical to the left half-site when the complementary strand is read the opposite way. Such palindromes allow a dimer of the same protein to bind to the exact same sequences. As can be seen in the preceding sequence comparisons, OR3 differs in one position (#4R), where a T-A base pairing has been substituted by C-G. Obviously, this substitution must account for the differential affinity of the repressor and Cro for OR1 and OR3.

The DNA-binding domain of the 434 repressor is made up of four helices that fold into a structure similar to other repressors, including the HTH motif (helices 2 and 3; the helix that is embedded in the major groove is helix 3). A fifth helix is necessary for subunit interactions because this repressor binds to DNA as a dimer (see Figures 5.17 and 5.21). Let us now examine the interaction of the 434 repressor dimer with OR1 and OR3. To do this, the structures of the 434 dimers with OR1 or OR3 will be superimposed. This can enable us to see what differences exist between the two models. In Figure 5.17, we can see the backbone traces of such interactions in stereo. The OR3 model is shown with solid lines, and the OR1 model has open lines. The upper half of DNA is the left half-site, and the lower half of DNA is the right half-site (compare with the

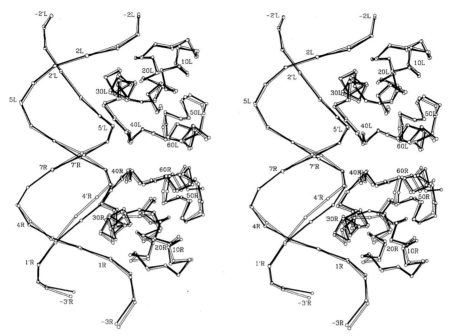

Figure 5.17. Interactions of the 434 repressor dimers with OR1 and OR3. The OR1/repressor model is shown with open lines, and the OR3/repressor model is represented by solid lines. The upper half of DNA is the left half-site (which is the same in OR1 and OR3), and the lower half of DNA is the right half-site (which differs in OR3 by T-A to C-G substitution). Note that the DNA backbones match very well in the upper half, while in the lower half there is a difference in the base pair 4. S. C. Harrison, Structure 1: 227–40 (1993). Reprinted with permission from Elsevier Science.

DNA sequence). It is obvious from this structural comparison that the DNA backbone at position 4′R bows out toward the protein. This bowing brings the phosphates closer to residues in the helix 2/helix 3 turn as well as to the amino terminus of helix 3. The protein monomer on the right half-site is slightly rotated resulting in decreased distance between the DNA backbone and residues of the turn. These structural features are shown in the Figure 5.17.

Let us now examine in detail the interactions at base pair 4. On the left half-site of OR1 or OR3 (consensus), a hydrogen bond links the O4 of thymine (4′L and 4′R for OR1) with the N-terminus of Gln-33 (of helix 3). Also, the methyl group of thymine is in van der Walls contact with the side chains of Gln-29 and Ser-30 (of helix 3) (Figures 5.18 and 5.19). The right site of OR1 interacts with the repressor in an identical manner as the left site of OR1. However, this is not true for the interaction of the repressor with the right site of OR3. The hydrogen bond between the substituted cytosine is there, but it is not direct. The hydrogen bond is mediated by a solvent; however, for mediation to occur, there must be a rotation of Gln-33. As a consequence of the rotation, the monomer shifts away from the DNA (about 1 Å relative to the left site). Also, the methyl

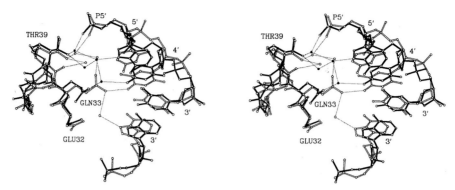

Figure 5.18. Stereo image of interactions between Gln-33 and OR1 (left half-site; open lines) and OR3 (right half-site; solid lines). S. C. Harrison, Structure 1: 227–40 (1993). Reprinted with permission from Elsevier Science.

group interactions are lost. These alterations result in weakened interactions of Gln-33 at base pair 4.

Alterations can also be seen in interactions between the phosphates and residues of the HTH motif of the 434 repressor. In the left half-site (consensus) Thr-26 and Thr-27 of the turn make hydrogen bonds with a water molecule, which in turn contacts the phosphate of 4′L. Ser-30 of helix 3 also interacts with the phosphate of 5′L via a water molecule. In the right half-site, where the DNA bows out, the interactions are different. Thr-27 and Ser-30 make direct contact with the phosphate at 4′R. As a result, the interacting residues rotate and come closer to the DNA backbone (Figure 5.19). Thus, it seems that interaction of the 434 repressor with the right half-site of OR3, which contains the C-G substitution, imposes some cost (because of the conformational change of the backbone). This cost should result in differential affinity of the repressor for the different ORs, with higher affinity for OR1 or OR2 (no cost) than OR3 (cost involved).

Why, then, does the Cro protein have a better affinity for OR3 than for OR1 or OR2? The Cro monomer is very similar to the repressor. The 3-D structure

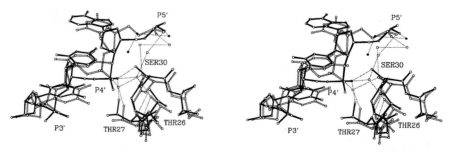

Figure 5.19. Stereo image of interactions of the HTH residues of the 434 repressor with the OR3 left (open lines) and OR3 right (solid lines) half sites. S. C. Harrison, Structure 1: 227–40 (1993). Reprinted with permission from Elsevier Science.

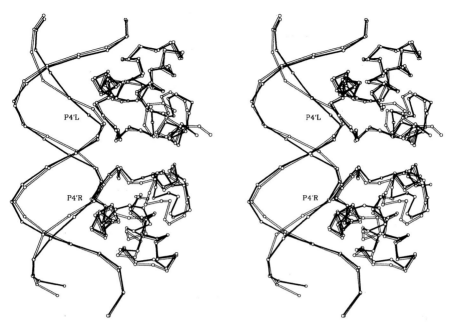

Figure 5.20. Comparison of OR1/Cro (open lines) with OR3/434 (solid lines). Compare with Figure 5.17 where the OR1/repressor and OR3/repressor are shown. S. C. Harrison, Structure 1: 227–40 (1993). Reprinted with permission from Elsevier Science.

of the DNA-binding domain is almost identical (compare with Figure 5.17). When the Cro dimer/OR1 is superimposed with 434/OR3, we discover that Cro is indifferent to the C-G substitution. In fact, the DNA backbone of OR3 assumes the same shift toward the protein in both the 4' left half-site (consensus) and the 4' right half-site (Figure 5.20). As a result, interactions at the 4'R position are very similar between Cro and the repressor. Because the Cro interaction with the left half-site does not involve any cost (the backbone adopts the same conformation as the nonconsensus right half-site), the interaction of Cro with OR3 is favored over OR1.

The Lambda Phage

Let us now present the structural characteristics of the proteins involved in the control of the lysogenic and lytic pathway in the lambda bacteriophage. The lambda repressor is encoded by the cI gene. The product of this gene turns off early transcription including transcription of Cro, establishing lysogeny. The lambda repressor occupies the OR1 and OR2 that, as in the case of the 434 bacteriophage, are part of the Cro promoter. Cro blocks OR3. When it does this, it turns off cI, and there is lysis. The competition here is between cI and Cro, but the winner is decided by another factor, the product of the cII gene. When CI is high, lysogeny is more likely, because CII activates transcription through a promoter that results in Cro antisense transcripts that block Cro transcription.

CII is protected against proteases by CIII, but a high concentration of proteases in the medium has a negative effect on CIII and destroys CII. Therefore, starvation favors lysogeny and rich media lysis.

The lambda repressor is made up of an N-terminus domain that possesses the DNA-binding activity and a C-terminus domain that is involved in dimerization. The N-terminus domain is made up of five helices with helices 2 and 3 being the HTH motif and helix 3 being the recognition helix. Helix 5 is involved in the dimerization. In Figure 5.21, the reader can compare the structures of lambda and 434 repressors bound to operators. The structure and the interaction are, in fact, very similar. This similarity in DNA binding is conferred by conserved amino acids found in the HTH motif of both repressors. The

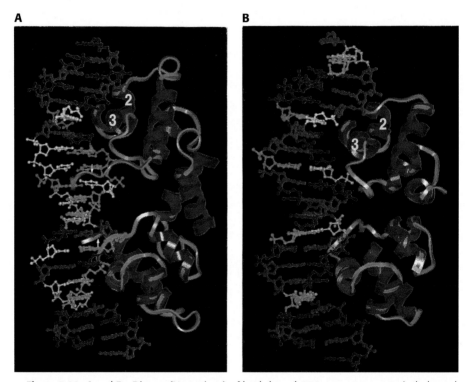

Figure 5.21. A and **B:** Dimers (N-terminus) of lambda and 434 repressor, respectively, bound to their operators. The helices are numbered with helix 3 embedded in the major groove. Also note that the lambda repressor has an arm that embraces the DNA (arrows). L. J. Beamer and C. O. Pabo, NDB file PRD010, J. Mol. Biol. 227: 177–96 (1992). L. J. W. Shimon and S. C. Harrison, NDB file PDR011, J. Mol. Biol. 232: 826–38 (1993). **C:** A stereo photograph of superimposed lambda and 434 half-sites. DNA is blue, and the HTH motif is yellow. Side chains for the conserved Gln-33, Gln-44, and Asn-52 that contact DNA are also shown (residue numbers for the lambda complex). C and D Pabo et al., Science 247: 1210–13 (1990). **D:** The interactions with a conserved A:T base pair. Hydrogen bonds are dotted lines. C and D reprinted with permission from American Association for the Advancement of Science.

C

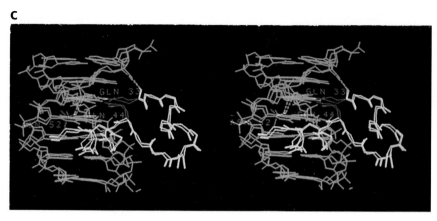

D

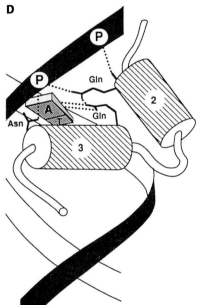

Figure 5.21(*cont.*).

amino acid sequences of these motifs follows, with bold letters indicating the conserved residues that make contact with the DNA and italics identifying the residues of the turn:

	33						44													52

Lambda: **Gln** Glu Ser Val Ala Asp Lys *Met Gly Met* Gly **Gln** Ser Gly Val Gly Ala Leu Phe **Asn**
434:     **Gln** Ala Glu Leu Ala Gln Lys *Val Gly Thr* Thr **Gln** Gln Ser Ile Glu Gln Leu Glu **Asn**

The lambda Cro has a structure similar to the repressor, but helices 4 and 5 have been substituted for beta strands in Cro. The relative orientation of Cro's helices 2 and 3 is virtually identical to that of the repressor. Let us see, however, how the repressor and Cro recognize the same operator sites, even though the repressor prefers OR1 and OR2 and Cro prefers OR3. As stated earlier, the recognition helix for lambda repressor is of the sequence GlnSerGlyValGlyAlaLeuPheAsn. The recognition helix of Cro is GlnSerAlaIleAsnLysAlaIleHis. They both begin from GlnSer but then they diverge. The interactions between these helices and the preferred operators follow:

Lambda repressor:	**Gln**	*Ser*	Gly	Val	Gly	*Ala*	Leu	Phe	Asn
OR1:	T	**A**	C	C	<u>T</u>	C	T	G	
	A	T	*G*	G	A	G	A	C	C
Cro:	**Gln**	*Ser*	Ala	Ile	<u>*Asn*</u>	<u>*Lys*</u>	Ala	Ile	His
OR3:	T	**A**	<u>T</u>	C	C	C	T	T	
	A	T	A	*G*	<u>*G*</u>	<u>*G*</u>	A	A	C

As can be deduced from these interactions, even though both the repressor and Cro recognize through the conserved Gln the A (both bold) and through the conserved Ser the G (both italics), the preference for OR1 or OR3 is determined by the interaction between the nonconserved amino acids of the recognition helix and the operators. For example, the repressor's Ala recognizes a T (both underlined), but Cro recognizes different bases by the nonconserved Asn and Lys (underlined italics).

TERMINATION OF PROKARYOTIC TRANSCRIPTION

Synthesis of RNA in prokaryotes is terminated by the formation of termination structures (mostly hairpins). Such structural formation is made possible by the sequence at the end of the transcript. In prokaryotes, we have two types of termination, the Rho-independent and the Rho-dependent termination (Figure 5.22). Rho is a factor that binds to the transcript as it is synthesized and literally chases the RNA polymerase up to the end of the transcript where the termination hairpin is formed. In Rho-independent termination, the hairpin is formed by inverted repeats characterized by G : C rich sequences that form a stem. Immediately after this stem, there is a sequence of Us. The formation of the hairpin and the pairing of Us with As (which form a weak double-stranded DNA : RNA hybrid) result in dissociation and the release of RNA polymerase and the end of transcription. In the Rho-dependent termination, the stem is not made of a G : C-rich sequence and the run of Us is also absent. Rho is a hexamer, and each monomer contains a domain with ATPase and helicase activity. However, the enzyme seems to function as a trimer of dimers. For example, out of the six potential high-affinity ATP-binding sites, there are three. Also, there are three strong nucleic acid-binding sites (preferring pyrimidines

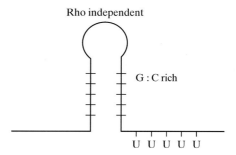

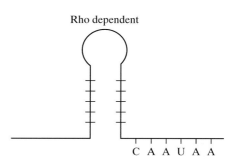

Figure 5.22. Termination hairpins in Rho-independent and Rho-dependent termination in prokaryotic transcription.

without DNA/RNA discrimination) and three weak sites (preferring C-rich RNA). The Rho ATP-binding site is homologous to F1 ATPases and is located in the two-thirds C-terminus of the enzyme. The N-terminal third contains the strong single-stranded nucleic acid-binding activity. This domain consists of a helical subdomain and a 5-stranded beta barrel, which belongs to the OB-fold class of single-stranded nucleic acid-binding motifs, such as the ones found in type II aminoacyl tRNA synthetase. The nucleic acid is bound to a cleft in the beta barrel. In Figure 5.23, we can observe the 3-D structure of the N-terminal third of Rho bound to oligoC. Note the cleft of the beta barrel that accommodates the RNA fragment. A closer view reveals the preference of Rho for pyrimidines (Figure 5.23B). In the structure, the first RNA cytosine is trapped by a wall created by Tyr-80, Glu-108, and Tyr-110. The base of the second nucleotide is stacked against Phe-64. A hydrophobic pocket for the sugar moieties is formed by Phe-62 and Leu-58. Such an arrangement cannot accommodate a purine, unless a large rearrangement of the loop is in place. Note the interaction between the RNA cytosine and Arg-66 and Asp-78 (Figure 5.23C). This interaction mimics considerably the interaction between a C and G (Figures 5.23D and 5.23E). Such an interaction reveals why Rho has preference for pyrimidines and not for purines. It is believed that the Rho helicase functions similarly to the hexameric helicases. However, the function of Rho helicase has been extended. Rho is able to translocate along the mRNA. Binding to the mRNA activates its ATPase, duplex unwinding, and translocation properties, which are necessary to terminate transcription.

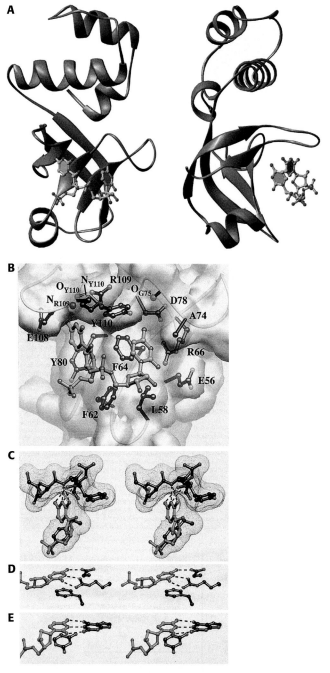

Figure 5.23. A: Two orientations of the N-terminal third of *E. coli* Rho with oligoC bound to the cleft of the beta barrel. **B:** Details of the interactions between Rho and oligoC. Cytosines are blue, and residues from Rho interacting with the Cs are brown. Note that C1 is locked by Tyr-80, Glu-108, and Tyr-110 and C2 is stacked against Phe-64. **C** and **D:** Hydrogen bonding between C2 and Arg-66 and Asp-78 and stacking against Phe-64. **E:** Interaction of a C with G with another C ring stacked against the interacting C. Note the similarity in the mode of interactions depicted in D and E. J. M. Berger, Mol. Cell 3: 487–93 (1999). Reprinted with permission from Elsevier Science.

The Attenuator

As indicated at the beginning of this section, another method of regulation of bacterial operons is by attenuation. Attenuation is employed because the regulation by repressor is rather weak. An example of such regulation is the *E. coli* tryptophan (*trp*) operon. This operon consists of five genes that are coded for the enzymes necessary for the synthesis of tryptophan. Control by attenuation is basically the formation of a termination hairpin before the genes of the operon are transcribed. Let us examine, however, the sequence of events involved in such regulation. The five genes of the *trp* operon are preceded by the leader and attenuator sequences. Four regions mark these sequences. Region 1 contains codons for tryptophan. Region 2 is complementary to regions 1 and 3, while region 3 is complementary to regions 2 and 4. In other words, pairing can occur between 1 and 2, 2 and 3, or 3 and 4 regions. Pairing of regions 3 and 4 results in a termination hairpin. When tryptophan is absent, the cell needs the enzymes to synthesize the amino acid. However, because tryptophan is scarce, the ribosome stalls considerably when it encounters the tryptophan codons. This allows 2 : 3 pairing, which disrupts the termination hairpin formation, and

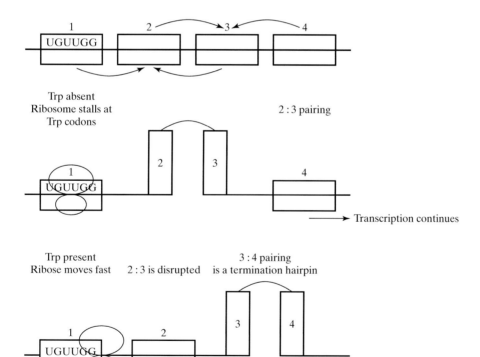

Figure 5.24. Structural features in the leader and attenuator sequences of the *E. coli trp* operon. Absence of tryptophan stalls the ribosome at tryptophan codons and complementary regions 2 and 3 pair. This action leads to continuation of the transcription. On the contrary, the presence of tryptophan results in the pairing of regions 3 and 4, which produces a termination hairpin.

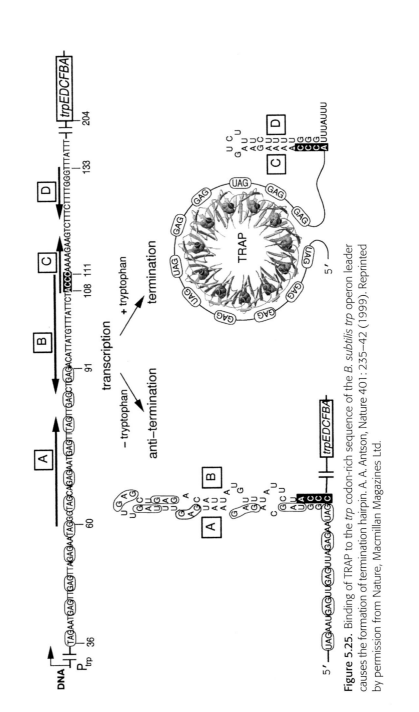

Figure 5.25. Binding of TRAP to the *trp* codon-rich sequence of the *B. subtilis trp* operon leader causes the formation of termination hairpin. A. A. Antson, Nature 401 : 235–42 (1999). Reprinted by permission from Nature, Macmillan Magazines Ltd.

transcription continues. When tryptophan is present, and there is no need for further synthesis, the ribosome moves rather quickly through the tryptophan codons and does not allow much time for 2 : 3 pairing, resulting, therefore, in the pairing between regions 3 and 4, which creates a termination hairpin (Figure 5.24). This type of regulation is very interesting because it involves sequences (codons) that are necessary for coding the very amino acid that is supposed to be synthesized. We can also find this type of regulation in operons that encode genes for the synthesis of other amino acids. In their leader sequence, all these operons contain codons for the amino acid whose synthesis they regulate.

In *Bacillus subtilis*, there is no *trp* repressor-controlling transcription, and regulation is achieved only by attenuation. Attenuation is not, however, controlled by the movement of the ribosome, as in *E. coli*, but rather by the *trp* RNA-binding attenuation protein (TRAP). Likewise in *E. coli*, the *B. subtilis trp* leader contains codons for tryptophan as well as similar regions. This protein has the ability to bind tryptophan as well as tryptophan codons. Therefore, when tryptophan is present, TRAP is able to bind the tryptophan codons. This causes downstream RNA to fold, which disrupts the pairing of regions A and B and allows the termination hairpin to form. In the absence of tryptophan, such RNA folding is not possible, regions A and B pair, and transcription proceeds.

TRAP is an 11-mer ring that can bind 11 tryptophan codons; therefore, by binding the downstream sequences of the leader, it can circularize the RNA that contains regions A and B that can form the antitermination hairpin (Figure 5.25).

CHAPTER SIX

Transcription in Eukaryotes

PRIMER Transcription in eukaryotes is more complicated than in prokaryotes. In eukaryotes, there must be tissue-specific gene transcription because many different cell types need to be created. Therefore, transcription must be selective in different cells and tissues. This selectivity is possible by the cell- or tissue-specific transcriptional factors. These factors can bind sequences present in promoters or enhancers of genes that will be specifically expressed in a particular cell type. The basal transcriptional machinery, which is assembled at the promoters of genes and is responsible for the recruitment of RNA polymerase, is also more complex in eukaryotes. In addition, in eukaryotes there are three RNA polymerases, each one transcribing a particular set of genes.

As in Chapter 5, I begin this chapter by presenting the 3-D structure of eukaryotic RNA polymerase II, which transcribes the genes that produce mRNAs. A comparison is made with the prokaryotic RNA polymerase. I then describe the factors that constitute the basal transcriptional machinery, their 3-D structure, and their interaction with DNA, RNA polymerase II, and each other. The reader should be able to obtain a very informative picture of this assembly that leads to the initiation of eukaryotic transcription. I then proceed to the regulation of transcription. As I mentioned earlier, regulation of transcription is unique for each of the more than 200 different cell types. Covering everything we know about regulation is, therefore, overwhelming. The goal here is to show how specificity is achieved, and to do this, only a few examples are necessary. These examples, however, demonstrate that the DNA-binding domains of tissue-specific regulators are few in number, including the helix-turn-helix motif and the zinc-binding domains, and that their specificities on DNA targets depend on the way they interact with DNA.

In the last part of this chapter, I present the 3-D structural characteristics of such DNA-binding motifs in an attempt to associate their structure with tissue-specific activation of genes. Again, this list is not exhaustive, but I am confident that it will effectively illustrate some of the basic mechanisms involved in tissue-specific regulation of gene expression.

Unlike the prokaryotes, where transcription is regulated by one RNA polymerase, eukaryotes three RNA polymerases regulate transcription. RNA polymerase I is responsible for transcribing the genes for ribosomal RNA (class I) and is located in the nucleoli. RNA pol II transcribes heterogeneous RNA (the precursors of mRNAs) and U RNAs (except U6), otherwise known as small nuclear RNAs (class II). RNA pol III transcribes the genes for tRNAs, 5S rRNA, U6, and some viral RNAs (class III). Detailed studies with all three RNA polymerases from yeast have revealed that 10 to 14 subunits can be found in all polymerases. All polymerases have two large subunits that are homologous in structure and function with the beta and beta' subunits of the prokaryotic RNA polymerase. Another smaller subunit is the equivalent of the prokaryotic alpha subunit. Apart from these subunits, all three polymerases share another five subunits. In RNA polymerase II, two subunits do not seem to play any role in transcription, and two others appear to be essential.

STRUCTURE AND FUNCTION OF THE YEAST RNA POL II

Most of the structural information concerning eukaryotic RNA polymerases has come by studying RNA pol II, especially the one from yeast. The yeast RNA pol II has 12 subunits, but the 3-D structure has been determined without 2 subunits (Rpb4 and Rpb7, which account for 8% of the mass, but their deletion does not seem to affect transcription and elongation). Of these ten subunits, nine are conserved in all I, II, and III. Let us briefly present some of these structural features and pinpoint some critical general similarities and differences with the bacterial counterpart. One of the major differences between the two large subunits of RNA pol II is that one of them, the IIa (Rpb1, homologous to bacterial beta') contains the peptide Tyr-Ser-Pro-Thr-Ser-Pro-Ser that is repeated many times. This peptide is found 26 times in yeast and 52 times in mouse. It is found in the C-terminus and is called the C-terminal domain (CTD). In electrophoretic studies, another larger subunit has been detected (IIo), but, in fact, IIo is the phosphorylated form of IIa. The repeated peptide of the CTD is phosphorylated. The unphosphorylated polymerase binds to the promoter (in this sense, it is equivalent to the prokaryotic RNA pol I holoenzyme) and is important in forming the initiation complex. The phosphorylated RNA pol II seems to function in the elongation process.

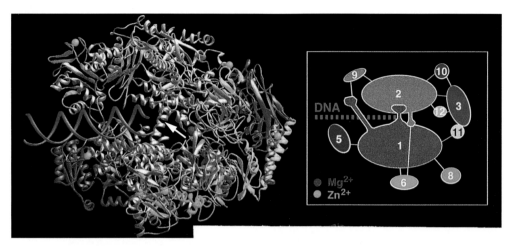

Figure 6.1. A model of the yeast 10-subunit RNA Pol II. The different subunits are represented with a different color code, which is explained in the insert. R. Kornberg, *Science* 288: 640–8 (2000). Reprinted with permission from American Association for the Advancement of Science.

The three-dimensional structure of the yeast RNA polymerase II has been solved and is remarkably similar to the prokaryotic counterpart. In Figure 6.1, a model of the 10-subunit yeast RNA pol II is presented. Obviously, the two large subunits (Rpb1 and Rpb2) constitute the claws and create a channel of 25 Å. This channel is formed when the tip of the arm is closed and makes a grip of the downstream DNA. The structures that grip the downstream DNA are the jaws made up from Rpb5 and regions of Rpb1 and Rpb9. A hinged domain (or the clamp) (N-terminal of Rpb1 and Rpb6 and C-terminal of Rpb2) is the equivalent of the flap seen in the prokaryotic RNA polymerase. An important compound of Rpb1 is the bridge helix (arrow in Figure 6.1), which plays a role in the incorporation of the incoming nucleotide.

Based on the length of the tunnel, it is estimated that it can accommodate 10 residues of single-stranded RNA. The hinged domain would function as a DNA clamp, and this might contribute to the great stability of eukaryotic transcription elongation complexes that can traverse one million base pairs before termination. The position of the residual (secondary) channel is also quite similar to the prokaryotic enzyme. During transcription, the DNA unwinds, creating a bubble where the DNA-RNA hybrid of about 8 to 9 bp with the 3′ end of RNA is located. Right there, the duplex DNA-RNA encounters a blocking element, the wall or flap. The arrangement of the clamp and the flap is important for the orientation of the hybrid and the path of the RNA transcript (Figure 6.2). The hybrid heteroduplex assumes a conformation, which is an intermediate between A- and B-DNA. The heteroduplex is also unwound. Abortive cycling yields transcripts of 2 or 3 nucleotides as well as transcripts up to 10 nucleotides. The bridge helix (pointed by an arrow in Figure 6.1) from

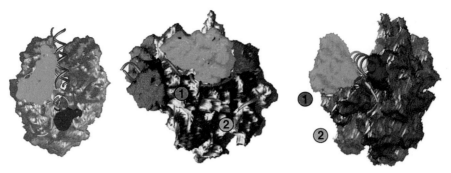

Figure 6.2. 3-D structure of the RNA polymerizing complex (three orientations). The RNA pol II is shown as a surface representation. The brown subunits are the jaws, green represents the clamp, and pink is the wall or flap. The rest of the polymerase is gray. Downstream DNA is blue and the hybrid at the active site is orange. Two pathways for the exiting RNA transcript are presented. Groove 1 (orange dashed lines) guides the transcript under the clamp near the N-terminus of Rpb1. Groove 2 is shown by dashed yellow lines. The hole near the hybrid at the active site is the equivalent of the prokaryotic residual channel that provides access to substrates. R. Kornberg, Science 288: 640–8 (2000). Reprinted with permission from American Association for the Advancement of Science.

Rpb1 plays a paramount role in the incorporation of NTP in the growing RNA transcript. The bridge helix binds the hybrid in such a way that allows the interaction of the NTP only with the +1 nucleotide of the substrate. When the incoming NTP is incorporated, changes in the conformation of the bridge helix allow for the translocation to position −1 and movement of nucleic acids to create the next empty +1 position, thus completing a cycle during RNA synthesis. At the upstream end of the hybrid (5′ end of RNA) the strands must separate. When the initiating complex reaches 10 nucleotides in length, the newly synthesized RNA separates from the DNA-RNA hybrid and enters an exit channel. For hybrid dissociation, parts of the clamp are important. The RNA strand enters a binding site extending from 10 to 20 nucleotides upstream of the active site. Two grooves in the enzyme seem suitable for providing the path to the RNA transcript (Figure 6.2). Groove 1, however, seems to be the choice for such a job. It has a length and location that can bind RNA 10 to 20 nucleotides from the active site. Also, this RNA path would lead back toward the downstream DNA close to the N-terminus of Rpb1. All this occurs in accordance with biochemical data cross-linking RNA to the N-terminus of the RNA polymerase. Finally, having the RNA transcript at the base of the clamp could explain the great stability of the enzyme. After the RNA exits from the polymerase, it should be available for processing, such as capping, which occurs when the transcript reaches a length of about 25 nucleotides. One fundamental difference between the eukaryotic and prokaryotic enzymes is that RNA pol II is not sensitive to termination hairpins. This difference might reflect the difference in the flap/hinged domain structure.

Overall, there are many structural similarities between the prokaryotic RNA polymerase and the eukaryotic RNA pol II. The most important difference, however, is the jaws, which are unique to the eukaryotic polymerase RNA pol II. This might reflect their interaction with trancription initiation factor TFIIE (see later). Also, the presence of the jaws might account for the protection of up to 20 bp of downstream DNA by RNA pol II, in contrast to about 13 bp, which are protected by the bacterial counterpart. As in bacteria, transcription begins with the repeated synthesis and release of short RNAs until it reaches the barrier of 10 nucleotides. This barrier is the point when the hybrid strands separate and the RNA transcript enters the exiting groove. When the transcript is about 20 nucleotides long we have a fully stable transcribing complex.

COMPARISON OF PROKARYOTIC AND EUKARYOTIC RNA POLYMERASES WITH THE T7 RNA POLYMERASE

As we discussed in the previous sections, the prokaryotic and the eukaryotic RNA polymerases are multisubunit enzymes. The T7 RNA polymerase from the T7 bacteriophage is capable of transcribing the entire T7 phage DNA. However, structurally the T7 RNA polymerase is closely related to the DNA polymerases (DNA pol I). Therefore, this enzyme is instructive in the study of what determines its specificity for ribonucleotide instead of deoxyribonucleotide. As noted earlier, DNA polymerases have a common overall structure, looking like a right hand with domains corresponding to thumb, fingers, and palm. The active site of the T7 RNA polymerase is located in a pocket created by the fingers, thumb, palm, and an N-terminal domain that is unique to the RNA polymerase. The active site differs from DNA pol I where the binding of the DNA is in a more open cleft. The O helix of DNA polymerases, which is important for binding the incoming nucleotide, contains a highly conserved sequence. However, the O helix of the T7 RNA polymerase differs in its structure. Tyr-639, which in other DNA polymerases is part of the helix, is located in a loop region adjacent to the O helix. This change is very important because it confers on the T7 RNA polymerase its specificity for ribonucleotides and not deoxyribonucleotides. Indeed, mutation of Tyr-639 to Phe-639 reduces the preference for ribonucleotides (Figure 6.3).

Let us now examine the synthesis of the RNA transcript in the active pocket. In Figure 6.4, we can examine the structure of the single-stranded template at initiation (A) and after synthesis of three nucleotides (B). Before the synthesis of the first $3' \rightarrow 5'$ phosphodiester bond, the template strand at position -1 binds a hydrophobic pocket (W422) very close to the active site. This brings template nucleotides $+1$ and $+2$ in a position that creates the binding sites for the two GTP substrates. After the synthesis of the pppGpGpG transcript, the template base -1 flips out of the small hydrophobic pocket W422 into the larger

A

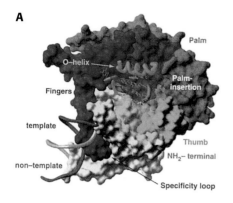

B

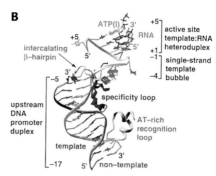

Figure 6.3. The structure of T7 RNA polymerase transcribing complex. The color codes are self-explanatory. **A:** The different domains are shown with template and nontemplate strands. **B:** The template and nontemplate strands viewed with the RNA transcript (green) and the next incoming ribonucleoside triphosphate (light green). **C:** The active site viewed with the pppGpGpG RNA transcript (green). Note that Y639 from the loop end of the O helix interacts with $T + 4$ and overlaps the position of the next incoming ATP. T. A. Steitz, Science 286: 2305–9 (1999). Reprinted with permission from American Association for the Advancement of Science.

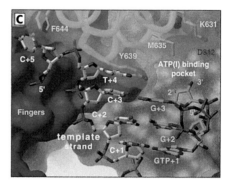

active site pocket. Also, the structure of the template residues $T{-}3$ to $T{-}1$ has changed. In Figure 6.4C we can see extension of the transcript. The transcript is exiting through an exit channel created by the unique N-terminal domain and the thumb. Such experimental data indicate that T7 RNA polymerase does not change conformations during the initiation of transcription.

These structural data also provide supporting evidence for DNA scrunching during initiation of transcription as opposed to other popular views of polymerase inchworming or back-and-forward DNA sliding. The volume of the active-site pocket indicates that it can accommodate the abortive product (8 to 10 bases) and 6- to 9-nucleotide-long template strands. The progressive phase starts when an additional template cannot be accommodated in the active site and when the transcript is long enough to bind tightly to the N-terminal domain. A longer transcript might induce conformational changes in the thumb that allows it to close and create the exit channel (Figure 6.4D). In this sense, the thump is equivalent to one of the claws of the prokaryotic and eukaryotic RNA polymerases.

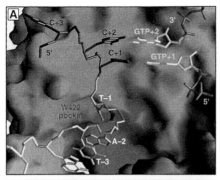

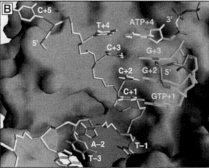

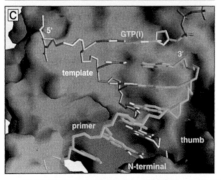

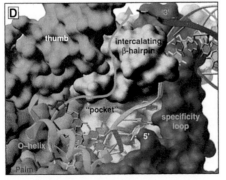

Figure 6.4. A, B, and **C:** Initiation of transcription by the T7 RNA polymerase. Note that the protein does not show any structural changes during the synthesis of the RNA transcript. Rather, the DNA template strand is structurally altered. **D:** : Progression of transcription and the exit channel. T. A. Steitz, Science 286: 2305–9 (1999). Reprinted with permission from American Association for the Advancement of Science.

REGULATION IN EUKARYOTES

We will examine regulation in eukaryotes according to the transcription by the different RNA polymerases. As mentioned at the beginning of this chapter, in eukaryotes different classes of genes are transcribed by different RNA polymerases.

Regulation in Class I Genes

Regulation in class I genes is regulation of ribosomal RNA genes, which are transcribed by RNA polymerase I. Let us first examine the organization of rRNA genes and their control elements. The rRNA genes are located in clusters, which are separated by the intergenic spacer (IGS). Each rRNA gene cluster contains the sequence for the 18S, 5.8S, and 28S rRNA in this order. The 18S is part of the small ribosomal subunit, and 28S and 5.8S are part of the large ribosomal subunit. These sequences are flanked by the 5′ external transcribed sequences (5′ ETS) and the 3′ external transcribed sequences (3′ ETS). The 5.8S is flanked by the 5′ and 3′ internal transcribed sequences (5′ and 3′ ITS). At the 5′ end of IGS, there are termination signals (which are reminiscent in sequence to the Rho-independent termination signal found in prokaryotes). At the 3′ end of the IGS and near the start of the rRNA gene cluster transcriptional unit, there are promoters. These class I promoters contain two elements the GC-rich core (from −45 to the start) and the upstream control element (UCE) (from −156 to −107 in human) (Figure 6.5).

Two transcriptional factors are involved in class I regulation. The upstream binding factor (UBF) binds the core element and UCE. SL1 is the other factor. It is composed of TATA-binding protein (TBP) and three TATA-associated factors (TAFs). SL1 does not bind to the rRNA promoter but interacts with RNA polymerase I to strengthen the binding of polymerase to UCE.

Termination of transcription is achieved by the termination signals found at the 5′ end of the intergenic sequences. In mouse, the termination sequence contains motifs that are inverted purine-rich followed by tracks of Ts. An example of one motif is AGGTCGACCAGTACTCCG, which can form a stem and is followed by an 18-mer containing 17 Ts. It is possible that a mechanism similar to the prokaryotic Rho-independent is in place.

Regulation in Class III Genes

As indicated earlier, class III genes are transcribed by RNA polymerase III and include the tRNA genes, the 5S rRNA gene, U6, and some viral RNAs. tRNA genes are organized in clusters. A cluster can contain the same tRNA gene in repeats (i.e., cluster of only tRNA^{Tyr}) or a cluster of different tRNA genes. 5S rRNA is produced by a different transcriptional unit and is part if the large ribosomal subunit.

The class III promoters are also simple and unique. Their uniqueness lies in the fact that the control elements are not upstream of the transcribed region

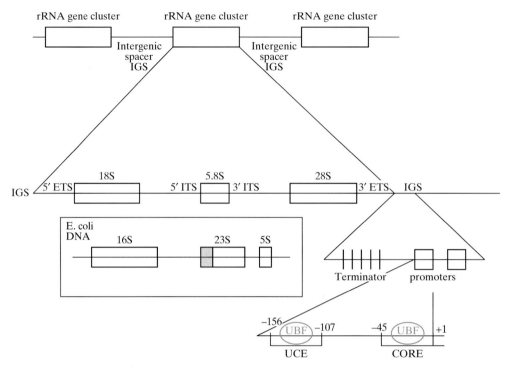

Figure 6.5. The organization of rRNA genes. Note the comparison with the rRNA gene cluster organization in *E. coli* in the insert. Here *E. coli* 16S and 23S rRNAs are followed by the 5S rRNA gene. The shaded region at the 5′ end of the 23S is homologous to the eukaryotic 5.8S rRNA in sequence and structure.

but are instead internal. The controlling region is called internal control region (ICR). In tRNA or adenovirus VA genes, the ICR is characterized by Box A and Box B (type 2 internal promoters). In 5S rRNA, the ICR is characterized by Box A, Box C, and an intermediate element (type 1 internal promoter). There are also some class III genes with pol II-like promoters. For example, the 7SL RNA gene contains a weak internal promoter as well as a sequence in the 5′ region of the gene that is necessary for high-level transcription. Other class III genes, such as 7SK and U6 RNA genes lack internal promoters and contain class II-like promoters with TATA boxes.

Several transcriptional factors assist RNA polymerase III in transcription. These factors are called transcriptional factors III (TFIII). All class III genes require TFIIIB and TFIIIC, with the exception of 5S rRNA genes, which also require TFIIIA. TFIIIC binds Box A and Box B. TFIIIB consists of TBP and two other proteins and binds to a region upstream the transcriptional start site. TBP is the TATA-binding protein (see later sections). After pol III interaction with TFIIIB, TFIIIC is displaced and pol III takes this place. TFIIIB remains bound in its original region ready to promote a new round of transcription. For

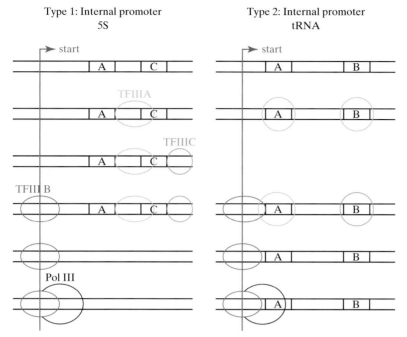

Figure 6.6. Sequential binding of transcription factors to type 1 and type 2 internal promoters in class III genes. A, B, and C are the boxes containing regulatory sequences in the ICR.

5S rRNA gene transcription, TFIIIA binds Box C first. Then TFIIIC binds at Box C, and TFIIIB binds at Box A with pol III (Figure 6.6).

Structure of TFIIIA

TFIIIA was the first eukaryotic transcriptional factor to be discovered. When the gene was sequenced, it provided interesting information that led to the identification of another very important DNA-binding motif: the zinc finger. The Xenopus TFIIIA protein contains nine repeated units that are quite homologous. The consensus sequence of these units was C-X_{2-4}-C-X_{12-13}-H-X_{3-4}-H. C is a cysteine, H is a histidine, and X is any other amino acid. Such a motif has the ability to bind zinc; the four ligands for this binding are provided by the two cysteines and the two histidines. In fact, the discovery of such DNA-binding domain sparked the search for zinc fingers in other proteins, which led to the discovery of numerous transcriptional factors (more about zinc fingers in the next sections). The three-dimensional structure of a TFIIIA-like zinc finger is characterized by two short antiparallel beta strands, containing the two cysteines, followed by an alpha helix that contains the two histidines (Figures 6.7 and 6.8). The alpha helix is the DNA-binding portion of the zinc finger. From the nine zinc fingers, the central group (fingers 4 to 7) binds to the 5S rRNA.

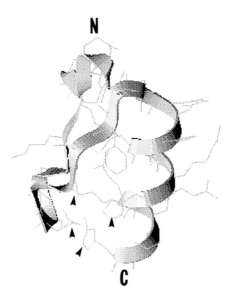

Figure 6.7. A model of a TFIIIA zinc finger indicating the antiparallel beta strands in the N-terminus (N), the DNA-binding helix in the C-terminus (C), and the four ligands for zinc (arrowheads). C. A. Kim and J. M. Berg, NDB file PDTB41, Nat. Struct. Biol. 3: 940–5 (1996).

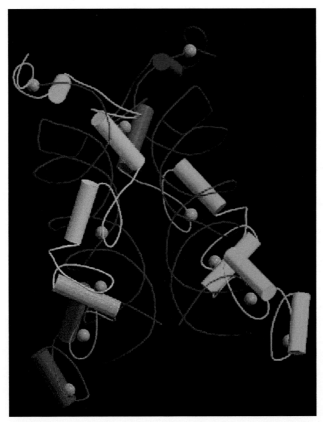

Figure 6.8. The N-terminal six fingers of Xenopus TFIIIA bound to 5S rRNA gene nucleotides +62–+93. R. T. Nolte et al., PDB file 1TF6, PNAS 95: 2938–43 (1998).

Regulation in Class II Genes

Regulation in class II genes is by far the most complex of the three classes. This complexity is characterized by more elaborate promoters and by a more complex basal machinery of transcriptional factors that in unison with RNA Pol II initiate the transcription of class II genes.

A typical class II promoter contains the TATA box at position −25, the initiator, an upstream element, and a downstream element. The initiator is the simplest form of a promoter, the required sequence being an A (the start of transcriptional unit) flanked by a pyrimidine at the 5′ and any nucleotide at the 3′. The upstream element could either be a GC-rich region or the CAAT box. The downstream element has no definite conserved or consensus sequences. Most of the class II promoters, however, lack some of these elements. Some have only the initiator, and others lack one of the upstream elements. Some housekeeping genes do not have a TATA box. In addition to these elements, class II promoters have elements that are unique to a particular gene or a particular set or group of genes. We will examine these unique elements later. At this point, we will continue with the basal transcriptional machinery required for class II promoter recognition and transcription.

The Basal Transcriptional Apparatus in Class II Genes

The factors that are involved in the basal transcriptional apparatus in class II genes will be presented based on their sequence of recruitment. Transcriptional factor IID (TFIID) is the first to occupy the TATA box. This factor is in fact made up of the TATA binding protein, TBP, and 8–10 TBP-associated factors (TAFs). TFIID is the principal TBP-containing complex, even though other complexes with TBP are known. Subsequently, transcriptional factor IIA (TFIIA) joins the complex. This factor might activate TBP by relieving a repression that is caused by TAFs. Then, TFIIB binds and protects the template strand from −10 to +10. The next factor, TFIIF, is made up of two subunits. The large subunit, called RAP74, has an ATP-dependent DNA helicase activity that could be involved in melting the DNA strands at the initiation. The small subunit, called RAP34, is the bacterial homolog of the sigma factor and binds RNA pol II. This factor is responsible for bringing RNA pol II to the complex. After TFIIF and polymerase binding, TFIIE binds and protects more downstream up to +30 and interacts with the jaws of RNA pol II. It also contains a subunit with a zinc finger DNA-binding domain. TFIIH has ATPase, helicase, and kinase activities and is associated with a protein kinase (MO15/CDK7), which is able to phosphorylate the CTD tail of RNA pol II. MO15/CDK7 could be a subunit of TFIIH, but it has been found free of TFIIH as well (see later). TFIIH is able, therefore, to activate RNA pol II to start transcription. Another factor, the TFIIJ, binds the complex and interacts with TFIIA. After the basal machinery has been assembled, the elongation process begins. For elongation, TFIIE and TFIIH are not needed, and they dissociate from the complex. However, elongation factors also exist. One of them is TFIIS, which promotes elongation by inhibiting long pauses. TFIIF also stimulates elongation. The preceding discussion leads the reader to assume that the preinitiation complex assembles at a class II promoter one factor at a time. This, however, might not be correct. There is evidence that the complex binds at a promoter as a preformed holoenzyme. The holoenzyme contains RNA polymerase II and a subset of the general factors presented earlier. Depending on the species, additional polypeptides might be part of the holoenzyme. In yeast, for example, these polypeptides are the SRB proteins. Let us now examine these factors, their 3-D structure, their interactions, and their function.

Structure and Function of TFIID. As indicated earlier, TFIID is composed of the TATA-binding protein and 8 to 10 TAFIIs. We call them TAFIIs because there are TAFs for both class I and class III. Different TAFIIs interact with promoter elements and help TFIID interact with specific transcriptional factors. TAFIIs are quite conserved, and they have a unique 3-D structure. They in fact fold similarly to the histone fold (Chapter 2). The fold is made up of three helices and a loop that create a surface for TBP interaction. In Figure 6.9, we can see a TAFII heterodimer and the interaction of a TAFII with TBP.

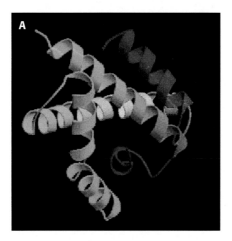

Figure 6.9. (**A**): The human TAFII28 (orange/yellow/green)/TAFII18 (blue) complex. Note the characteristic histone fold. C. Brick et al., PDB file 1BH8, Cell 94: 239–49 (1998). (**B**): The TBP-TAFII230 (from *Drosophila*) complex in two orientations. Arrowheads indicate the N- and C-terminal stirrup. M. Ikura, Cell 94: 573–83 (1998). Reprinted with permission from Elsevier Science.

TBP

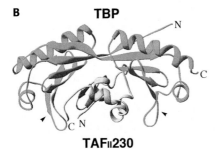

TAFII230

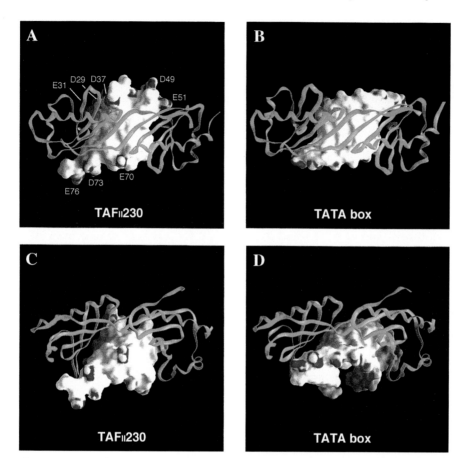

Figure 6.10. **A** and **C:** Two different orientations showing the interactions of TBP (green) with TAFII230 (surface representation). **B** and **D:** Same orientation as in A and C, respectively, but here TBP interacts with a partially unwound TATA box. Note the similarity between the surface interaction between TAFII230 and the TATA box. M. Ikura, Cell 94: 573–83 1998. Reprinted with permission from Elsevier Science.

TBP has a saddle-like structure, containing 10-strand beta sheet that forms a hydrophobic concave surface that lines up with the DNA. The beta sheet is flanked by two helices on the top convex surface that can interact with other factors (Figure 6.9). TBP contains specific sequences called the N-stirrup and C-stirrup, which mediate interactions with other factors (arrowheads in Figure 6.9). The underside of TBP interacts with the minor groove, opens it, and bends the TATA box nearly 80 degrees (Figure 6.10). In some cases, binding TAFs to TBP can inhibit binding of TBP to the TATA box. Examining the structural interactions of TBP with the TATA box and TBP with TAFIIs has revealed an interesting complementarity, which can explain how TAFs can exert inhibitory effects of TBP binding to the TATA box. (Figure 6.10). In *Drosophila*, the N-terminal of TAFII230 binds to TBP and inhibits TBP binding to the TATA box.

The mechanism of this inhibition lies in the structure of the TAFII230 because its surface that interacts with TBP is very similar to the surface of a partially unwound TATA box. In other words, it can compete, by molecular mimicry, with the TATA box for the binding to TBP. This structural mimicry can be clearly seen in Figure 6.10.

Structure of TFIIA. After TBP recognizes the TATA box, TFIIA binds the complex. TFIIA stabilizes the interaction between TBP and the TATA box by interacting with TBP and DNA directly. Even though it seems that TFIIA is not required for TBP-directed transcription in vitro, it does play an essential role in vivo. TFIIA is shaped as a boot, with two domains, one made up of a 12-strand beta barrel (the leg of the boot) and the other of four alpha helices (the foot of the boot). The binding interface between TFIIA and TBP (in yeast) is a beta strand of TFIIA and a beta strand (N-terminal stirrup) from TBP. This binding results in a continuous 16-strand beta sheet that curves across the underside of TBP bound to DNA. The rest of the beta sheet is the exposed surface of the TFIIA beta barrel (Figure 6.11). The beta barrel contacts DNA over 4 or 5 bases upstream the TATA box. It seems that TFIIA binds DNA only in the presence of TBP due to the resulting distortion in the DNA after TBP binding.

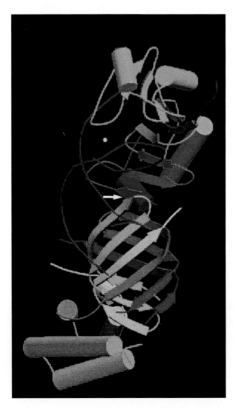

Figure 6.11. Structure of yeast TFIIA bound to yeast TBP and DNA. Note the interface between TBP and TFIIA (arrow) and the creation of the continuous 16-strand beta sheet. Also note the bending of the minor groove due to TBP binding (asterisk) and the four free helices of TFIIA that do not interact with DNA. Tan et al., PDB file 1YTF, Nature 381:127 (1996).

Structure of TFIIB. TFIIB consists of two related conserved domains and binds to DNA in a very specific manner. One domain (C-terminal) can recognize and interact with the major groove of the upstream sequence to the TATA box, and the other (N-terminal) interacts with the minor groove of the downstream sequence to the TATA box. This kind of interaction is only possible because of the bend in DNA induced by the binding of TBP to the TATA box. By all means, this interaction with DNA is very informative. Biochemical studies have shown that TBP recognizes the TATA box in both orientations, but this recognition cannot account for the unidirectionality of the preinitiation complex. Obviously, this asymmetry of TFIIB binding the upstream and downstream sequences to the TATA box can account for the unidirectionality of the assembly complex. Further, this recognition indicates that the distortion of the DNA sequence by TBP binding is required for TFIIB binding. The C-terminal domain of TFIIB and the C-terminal stirrup of TBP mediate the interaction between TBP and TFIIB. TFIIB interacts with the upstream major groove by a helix-turn-helix motif in a manner similar to that described for bacterial DNA-binding protein (see Chapter 5). Interaction with the downstream minor groove is mediated by a loop. TFIIB's downstream location (N-terminal domain) allows it to interact with RNA polymerase II and/or TFIIF and to link them to TFIID (Figure 6.12A). The relative position of TFIIB to RNA polymerase

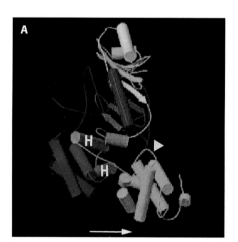
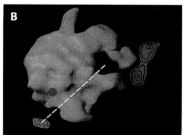

Figure 6.12. A: Interactions of human TBP, TFIIB, and DNA. TBP with its concave beta-sheet surface interacts with the TATA box (red) and distorts the minor groove. TFIIB interacts with its C-terminal domain (blue with the HTH light blue) and the C-terminal stirrup of TBP (orange). TFIIB interacts with the downstream sequence through a loop (arrowhead), which belongs to the N-terminal domain (turquoise and green). An arrow indicates the direction of transcription. D. B. Nikolov et al., PDB file 1VOL, Nature 377: 119 (1995). **B:** Position of TFIIB and TFIIE in relation to the yeast RNA polymerase II. TFIIB is green, TFIIE is magenta, and RNA polymerase II is blue. The red dot indicates the location of CTD, and the broken yellow line identifies the path of the DNA. R. Kornberg, Cell 85: 773–9 (1996). B reprinted with permission from Elsevier Science.

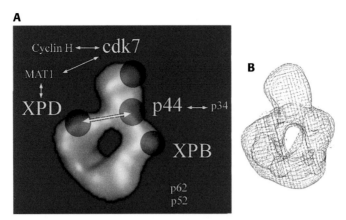

Figure 6.13. A: Quaternary organization of human TFIIH. The red dots indicate the location of cdk7, XPD, XPB, and p44 as determined by immunolabeling with specific antibodies. The arrows indicate the possible location of polypeptides as inferred by established subunit-subunit biochemical interactions. **B:** Superimposition of the TFIIH (blue) with a structure lacking the CAK complex, (red) to indicate the location of CAK complex (red). P. Schultz, Cell 102: 599–607 (2000). Reprinted with permission from Elsevier Science.

and TFIIE has been revealed by crystallography. It seems that TFIIB's (TATA-binding domain) distance from the active site of RNA polymerase II is about 110 Å, corresponding to a length of 32 bp, which is close to the distance of the TATA box from the transcription start site. This distance might indicate that TFIIB brings DNA in contact with RNA polymerase II by binding both TBP and the TATA box and thus defining the start site. TFIIE, on the other hand, is associated with the arms of RNA polymerase II and may play a role in stabilizing the closed state (Figure 6.12B).

Structure of TFIIH. As mentioned earlier, TFIIH has ATPase, helicase, and kinase activities. TFIIH can be divided into two major complexes. The core contains two ATP-dependent DNA helicases that act in opposite polarity, the XPB and XPD and the p62, p52, p44, and p34 polypeptides. The other complex (the cdk-activating kinase, CAK) is composed of the cdk7 kinase, cyclin H, and MAT1. Electron microscopy and image processing analysis of the human TFIIH has shown that it is ring-shaped, with the CAK complex protruding from the ring (Figure 6.13).

TRANSCRIPTIONAL ACTIVATION IN EUKARYOTES

The factors mentioned in the preceding section constitute the basal transcriptional apparatus; they are all general and responsible only for low basal levels of transcription. However, genes can be activated and achieve very high levels of transcription with the employment of gene-specific transcriptional factors, or activators. These factors recognize elements in the promoter or other regions of the gene and, by binding to them, are able to recruit the general factors and enhance the levels of transcription considerably. These factors, therefore, are specific for the gene(s) they are able to activate. The importance of these factors becomes greater when we think about tissue-specific gene transcription. Because these factors are gene specific, they can also activate genes in a tissue-specific manner. The eukaryotic body contains many different cell types (in humans there are more than 200); consequently, such specific gene regulation becomes imperative for establishing the identity of the different cell types and tissues during development and throughout life. These activators bind elements that are present in the 5′ regulatory region near the promoter or in other regions of the gene called enhancers. Enhancers can be found anywhere in the gene – in places such as introns and 5′ regions far away from the promoter. The elements of the enhancers, therefore, do not have to be near the transcription start site or in a particular orientation as the main promoter elements. How then can the activators interact with the elements of the promoters from far away? DNA looping is the answer. Obviously, activators can interact with proteins bound to the promoter and by doing so DNA must be looped. Architectural proteins, such as the ones mentioned in Chapter 2, help in this process.

Transcriptional activators posses at least two important domains. One is the DNA-binding domain and another is the transcription-activating domain. The transcription-activating domain can be acidic, glutamine-rich, or proline-rich. Also, most of them contain domains that allow them to dimerize. To date, many activators have been identified, and their mode of interaction with DNA has been revealed at the 3-D level. Their structure has provided very crucial information about their specificity in gene regulation and their function in cell fate and tissue determination. Within the scope of this book, we will provide a 3-D perspective of gene regulation and probe the molecular mechanisms involved. Therefore, it is important to examine the structure and function of these activators. I do not, however, wish to overwhelm the reader by presenting too much information on activators. I have selected a few examples to study specificity of gene regulation and the activators involved. This selection will enable us to study the most basic structural features involved in DNA binding and regulation.

Liver-Specific Transcription

As is true for most genes, liver-specific genes do posses the basal elements in their promoter. These are the TATA box (for mouse albumin gene at -28) and the CAAT box (at -82). However, in the albumin 5′ region, other elements can affect transcription. One of them is located at -59 to -47 and is the palindromic sequence GTTAATTAATCTAC. Absence of this element can affect transcription of the albumin gene by 50×. This sequence is the binding site of a liver-specific activator, the hepatocyte nuclear factor-1 (HNF-1). Other elements upstream (up to -160) can affect transcription by 5 to 15×. The HNF-1 binding site is found in other liver-specific genes from many species. The location is not always the same in different genes and can be found as far as 120 bases from the transcription start site. The HNF-1 binding site is the consensus sequence:

$$5'\text{-G T T A A T N A T T A A C-}3'$$
$$3'\text{-C A A T T A N T A A T T G-}5'$$

This is a palindrome; the sequence is the same when it is read on the opposite strands. This configuration is important because it alludes to the fact that proteins bind as dimers (see also the section entitled "Regulation in Prokaryotes" in Chapter 5).

The GAL Regulatory System

In prokaryotes, the galactose operon consists of three genes that encode for the necessary enzymes to metabolize galactose to glucose. As we know, all these genes are regulated together by the action of the GAL repressor. In eukaryotes (yeast), such a metabolic pathway is controlled by six genes, which are located in different chromosomes. Consequently, all six genes are regulated separately. GAL2 is located on chromosome XII and encodes a permease that allows exterior galactose to enter the cells. GAL1 encodes a galactokinase that converts galactose to galactose-1-phosphate. GAL7 encodes a galactose-1-phosphate uridyl transferase that forms uridine diphosphogalactose (UDPGal) from galactose-1-phosphate and uridine diphosphoglucose (UDPG). GAL10 encodes an epimerase that regenerates UDPG from UDPGal. GAL1, GAL7, and GAL10 are all located on chromosome II. When galactose is present, an inducer derived from galactose causes dissociation of the inactive complex made up of GAL4 and GAL80. GAL4 is located on chromosome XVI, and GAL80 is on chromosome XIII. GAL4 protein is the activator and binds to a specific sequence in the 5' upstream region of all the other GAL genes. The sequences are called upstream activating sequences (UASs). UASs are repeated several times in some of them (four times in GAL1 and GAL10) and can be found anywhere between −100 to −350. The consensus sequence of UASs is also a palindrome:

$$5'\text{-C G G A G/C G A C A G T C G/C T C C G-}3'$$
$$3'\text{-G C C T C/G C T G T C A G C/G A G G C-}5'$$

GAL4 protein contains its DNA-binding domain near the N-terminus and is a member of the zinc-containing family of DNA-binding proteins. We have already encountered zinc fingers in TFIIIA, but the different zinc-containing binding domains found in class II gene activators will be discussed in detail in a later section. The transcription-activating domain is a 49-residue domain with 11 acidic amino acids. Two monomers can dimerize by forming a parallel-coiled coil using two alpha helices (see the section entitled "The Zinc-Binding Domains" later in this chapter).

The Steroid Hormone Receptors

Steroids are lipids without glycerol, and their basic structure is four interlocking rings of carbon atoms. The precursor of all steroidal hormones is cholesterol, and differences in the biological activity between the different steroidal hormones are the result of small alterations or substitutions in the cholesterol backbone. Examples of steroidal hormones are estrogen, glucocorticoids, mineralocorticoids, androgens, and vitamin D. These hormones can bind the so-called nuclear steroid receptors in the nucleus. Receptors similar to the steroidal ones are the thyroid receptor, which binds the thyroid hormone (whose precursor is tyrosine), and the retinoid receptors, which bind the active metabolites of vitamin A (whose precursor is the carotinoids in plants and retinyl esters in animal fat). The superfamily of these receptors is divided in two classes. Thyroid hormone, retinoic acid, and vitamin D receptors belong to class 1, and the rest of the steroid hormone receptors belong to class 2. These receptors, once activated by the binding of their ligand, dimerize and bind to the so-called hormone responsive elements (HRE), which are found in the enhancers of responsive genes. For some ligands, there is only one receptor. For example, there is only one receptor for the different vitamin D metabolites. For other ligands, such as vitamin A metabolites, there are many receptors. For example, all-trans retinoic acid binds different isoforms of retinoic acid receptors (RARs), while 9-cis retinoic acid is the ligand for a different receptor, called RXR. In fact, RXR not only can homodimerize with retinoid X receptor (RXR), but it also heterodimerizes with other receptors, such as the vitamin D receptor (VDR) and thyroid receptor (TR). The discovery of these receptors shed light on the mechanism of gene activation by steroid hormones. As it was already known, these hormones can elicit a broad spectrum of effects in the organism. Vitamin A, for example, has diverse effects. It can affect the skin, it can induce cancer, it can cure cancer, it affects the nervous system, and it is instrumental for the morphogenetic program during development. This range of effects can be explained by identifying the nuclear receptors for the different metabolites of vitamin A. Different RARs can bind to different responsive elements and activate different set of genes that can affect different tissues.

Therefore, except the DNA-binding domain and the transcription-activating domain, the steroid and thyroid receptors also have a ligand-binding domain, which is present in the C-terminus of the activators. The ligand-binding domain is very specific and can discriminate even when similar metabolites are bound. The DNA-binding region of these nuclear receptors is made up of two zinc-binding domains. The first zinc finger controls specificity of DNA binding, and the second controls specificity of dimerization. The zinc-binding domains differ in structure from the ones found in TFIIIA or in GAL4. The differences in structure and function will be presented in a following section.

Specificity of the DNA-binding domain depends on the HREs, which are different for every steroid receptor. For example, the glucocortocoid responsive elements have the following consensus sequence, which is a palindrome separated by three nucleotides:

$$5'\text{-G G T A C A N N N T G T T C T-}3'$$
$$3'\text{-C C A T G T N N N A C A A G A-}5'$$

The thyroid responsive element in rat growth hormone gene is somewhat similar, but the palindromic sequences are separated by six nucleotides:

$$5'\text{-G A T C A N N N N N N T G A C C-}3'$$
$$3'\text{-C T A G T N N N N N N A C T G G-}5'$$

However, when THR, VDR, or RAR heterodimerize with RXR, they all recognize a palindromic sequence consisting of inverted repeats of the sequence AGGTCA. The specificity of these heterodimers is mediated by a second type of response elements consisting of direct repeats of the sequence AGGTCA separated by a nonspecific sequence of a variable length. For example, the direct repeat AGGTCA is separated by 3 nucleotides for the VDR-RXR heterodimer, by 4 for THR-RXR, and by 5 nucleotides for RAR-RXR. Also, specificity of a receptor for HRE could rest in the amino acid sequence of the DNA-binding domain. For example, the change of two amino acids (Gly and Ser to Glu and Gly) found at the base of the first finger can account for discrimination between glucocorticoid and estrogen responsive elements. The 3-D structure of zinc fingers and their interaction with HRE will be presented later, and it will help us visualize these specificities.

The Homeodomain-Containing Genes

So far, a few examples have been presented to show the degree of specific gene regulation in eukaryotes. The reader, no doubt, has realized how much more complex regulation of eukaryotic class II genes is, when compared with prokaryotic or even eukaryotic class I and III gene regulation. But let us for a moment marvel about the complexity of embryonic development. The embryo starts as a zygotic cell, which divides to form the blastula. Then, determination of three different tissues – ectoderm, mesoderm, and endoderm – marks the critical stage of gastrulation. These three tissues will eventually give rise, by differentiation, to all different cell types (more than 200) found in the animal body. The creation of all these cell types is possible through an amazing orchestration of inductive mechanisms. It has long been thought that master genes are the conductors that control this well-orchestrated sequence of events. And because induction must turn on sets of genes, unique transcriptional activators must be involved. In 1984, the first such activators were identified. For the history, these activators are coded by genes that control pattern formation and segmentation during *Drosophila* development. The genes were Ultrabithorax (a mutation in this gene causes duplication of the whole thorax structures), fushi tarazu (meaning less segments in Japanese; a mutation in this gene results in embryos with less segments), and antennapedia (a mutation in this gene transforms the antenna into a leg). As expected, these proteins are nuclear and have DNA-binding properties. The DNA-binding domain is a helix-turn-helix motif similar to the ones found in bacterial repressors. The domain was named homeobox or homeodomain because it was found in homeotic genes, and the genes containing them are now called Hox genes. The discovery of homeobox opened new avenues in the study of gene regulation during development. It was soon found that there is a cluster of homeobox-containing genes in *Drosophila* and that they are expressed sequentially in the animal body in the order they are positioned in the cluster. For example, the most 3′ Hox gene is expressed in the anterior part of the body while the most 5′ is expressed in the posterior part of the body. What is most fascinating, however, is that a similar arrangement of Hox genes and their expression occurs in higher vertebrates and mammals. In higher animals, the cluster was duplicated two times, and there are four clusters of Hox genes A, B, C and D. As in *Drosophila*, expression of the 3′ Hox genes demarcates the more anterior parts of the body, whereas expression of the 5′ Hox genes identifies the posterior parts (Figure 6.14). This colinearity between location of genes in the chromosome and expression along the body axis is unprecedented (just remember that the few genes involved in a simple metabolic pathway, such as the GAL genes, are located on different chromosomes). It indicates how the sequential expression of Hox genes might set the cascade of events that build up the body structures. It also indicates that the molecular mechanisms involved in making the body plan are highly conserved in the animal kingdom.

An excellent example of this type of action in Hox genes is the expression and involvement of the 5′ Hox genes from the A, C, and D cluster during limb development (the B cluster does not contain Hox 10 to 13 genes). During limb development, the 5′ genes are expressed more posteriously and distally, with HoxD13 being the first. Such expression specifies the most distal structure of the limb (fingers). Then, HoxD12, -11, -10, and -9 appear sequentially and mark more anterior and proximal regions, specifying more proximal structures of the limb. When mice lacking these genes were generated, such expression was translated to function. Animals lacking any of the Hox 13 genes developed with defective distal structures. Animals lacking the A11 and D11 genes developed without ulna and radius, and animals without C10 or D10 developed with defects in the humerus. In other words, as the genes are lost in the 5′ to 3′ direction, defects

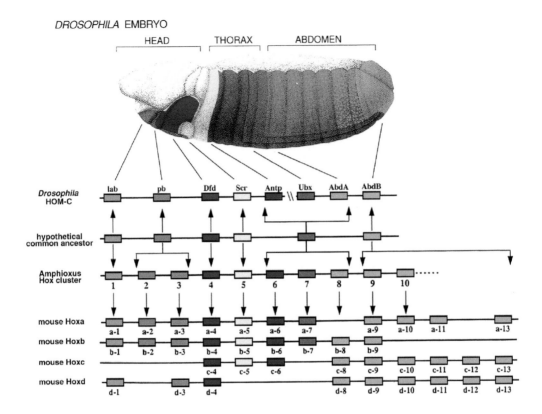

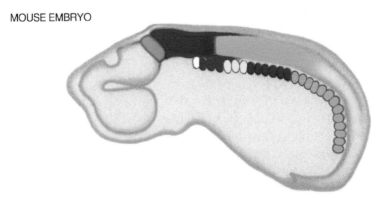

Figure 6.14. Location and expression along the anterior-posterior axis of Hox genes in *Drosophila* (top panel). The one cluster was duplicated after Amphioxus. In mouse and the rest of mammals there are four clusters. The Hox genes in the four clusters are also expressed in the mouse embryo (bottom panel) in the same order as their position in the chromosome, that is the most 5' genes (green) are expressed in the posterior and distal parts, while the most 3' genes (red, blue) are expressed in the anterior and proximal parts. Courtesy of Dr. S. Carroll.

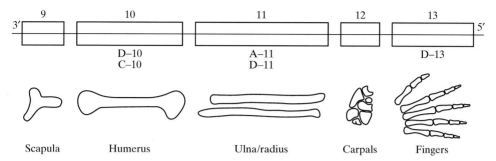

Figure 6.15. Hox genes specify structures during limb development. The 5′ Hox genes specify the most distal structures, and the 3′ Hox genes designate the more proximal structures.

were seen in the distal to proximal direction. Conceptually, we can imagine that loosing all Hox 9 to 13 genes in all clusters would result in a limbless animal (Figure 6.15).

The homeodomain is composed of 60 amino acids and is astonishingly conserved in the animal kingdom. Even though the rest of a Hox gene is not that homologous among species, its homeobox can be 100 percent identical. The 3-D structure of the homeodomain will be presented in detail in following sections, but some structural features are discussed now. The secondary structure of the homeodomain suggests that its structure is dominated by three alpha helices. The N-terminus of helix 1 contacts the minor groove. Helices 2 and 3 are in the HTH motif, with helix 3 contacting the major groove of DNA. Helix 1 is also very conserved among homeoboxes and might be involved in dimerization. Helix 3 contains the most conserved sequence found in homeoboxes. In fact, two amino acids in helix 3 – tryptophan and phenylalanine – are almost absolutely conserved and have become markers for identifying divergent homeoboxes. The homeoboxes bind 5′ regulatory elements and enhancers with preference for sequences containing TAAT.

The Mediator

We have examined a few examples of specific transcriptional regulation by activators. As previously mentioned, most of these factors bind elements that are no closer to the promoter than the RNA polymerase II holoenzyme. Previously, it was suggested that these factors come in contact with the basal machinery by DNA bending. In fact, an interaction between the activator and other proteins that interact with RNA polymerase II holoenzyme could induce such a bending. These proteins comprise the mediator, which seems to interact with RNA polymerase II via the CTD. The interaction of mediator and RNA polymerase II has been revealed by electron microscopy (Figure 6.16), and it seems that its structure is conserved among eukaryotes.

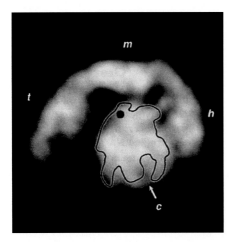

Figure 6.16. Interaction of the yeast mediator with RNA polymerase II. Mediator is presented here with three distinguishable domains: the tail (t), the middle (m) and head (h). RNA polymerase II is outlined, and the CTD is the black dot. R. Kornberg, Science 283: 985–7 (1999). Reprinted with permission from American Association for the Advancement of Science.

High levels of transcription, like those already described, depend on high levels of initiation of transcription as well as of reinitiation. We have described the important factors necessary for the initiation: the basal machinery and the activators. After initiation, a subset of factors such as RNA pol II and TFIIB dissociate from the complex. To reinitiate transcription, these factors must be recruited again. In yeast, this step requires the cooperative binding of TFIIB, RNA pol II, and the mediator.

CHROMATIN STRUCTURE AND GENE REGULATION

When we look at electron micrographs of the nucleus, we can see that some areas are darker than others. This difference is based on the degree of condensation. The darker areas are heterochromatin, which is condensed DNA throughout the cell cycle. The light areas are euchromatin and is decondensed DNA in interphase. The euchromatic regions (genes and unique sequences) replicate in early S phase, while heterochromatic regions (mostly repetitive regions) replicate late in S phase. In the past few years, however, researchers have realized that heterochromatin can affect expression of euchromatic regions. Stochastic heterochromatinization (juxtaposing euchromatin to heterochromatin) can result in inherited gene silencing. This is called position-effect variegation (PEV). Much about this phenomenon was studied with transgenic mice where, depending on the integration site, the gene would be expressed or not. Studies with such transgenic animals revealed that some genes have regions that contain cis-elements called locus control regions (LCRs), which initiate or maintain (or both) a cell-type-specific open chromatin (euchromatin) structure. One example of PEV is the expression of the gene that controls the red color in the eye of *Drosophila*. In mutants with brown eyes, it was found that this gene was not mutated, but it was silenced by heterochromatic regions that

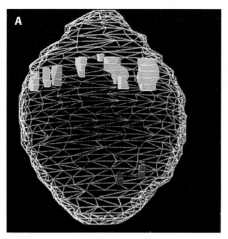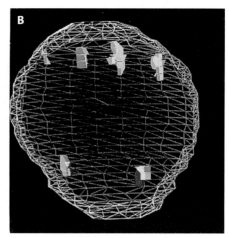

Figure 6.17. A: The *brown* gene (red) required for normal-colored eyes in *Drosophila* is localized far away from heterochromatic sequences (green) in the wild type. **B:** In the mutant white eye, the gene is co-localized with heterochromatic sequences. J. W. Sedat, Cell 85: 745–59 (1996). Reprinted with permission from Elsevier Science.

were juxtaposed near it (Figure 6.17). Another example of gene inactivation by condensation is the X-chromosome inactivation.

As we have already seen, DNA is packed mainly by histones, so any alterations in the chromatin structure that might alter gene expression should involve them. Indeed, a particular chemical modification in histones seems to be paramount for activation of genes. The chemical modification is the addition of acetyl groups in histones. Such an acetylation makes the histones loosen their grip on DNA so that a repressed gene can be activated (Figure 6.18). Interestingly, the players involved in this kind of gene activation include retinoid receptors. The inactivated retinoid receptor RAR/RXR heterodimer is bound by repressors and histone deacetylase, which removes acetyl groups from

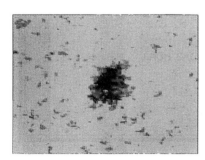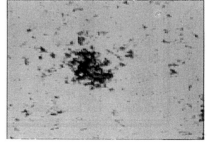

Figure 6.18. Compact unacetylated nucleosomes (left) with DNA (red) tightly wound around the core of histones. Acetylation causes spreading of the nucleosome and allows DNA to elongate (right). D. P. Bazett-Jones, Nucleic Acids Res. 24: 321–30 (1996). Reprinted with permission from Oxford University Press.

histones, resulting in tight DNA wrap. When retinoic acid binds the dimer, the activated receptors displace the repressors and recruit acetyltransferases. A particular TAFII has also been found to have acetylase activity. Acetylation then loosens up the tight association between histones and DNA, allowing binding of the basal transcriptional factors.

It seems that this kind of transcriptional control could be very important. It is interesting to speculate here that such a control might account for the silencing of whole clusters, such as Hox clusters. This control might also explain anatomical changes (such as the loss of limbs), which otherwise would require mutations in many genes and the possible loss of these genes in evolution, which is not the case.

REGULATION OF CELL FATE BY HOMEODOMAINS; THE YEAST MAT GENES

Yeast exists as three cell types – two haploid and one diploid. One of the haploid cells, the **a** cells, carry the MAT**a** allele and express **a**-specific genes. The other haploid cells, the α, carry the MATα allele and have **a**-specific genes turned off, while a new set of genes, the α-specific genes, are turned on. Haploid **a** and α cells can mate and form the third cell type, the **a**/α diploid cell. For such a regulation, transcriptional repression is very important. In both α and **a**/α, the MAT α2 gene product with the ubiquitous transcription factor MCM1 form a heterodimeric repressor that binds **a**-specific gene operators and turns off **a**-specific genes. In diploid cells, however, α2 binds with the MAT**a** gene product **a1** to the haploid-specific gene operator; as a result, haploid-specific genes are repressed. **a1** does not bind to DNA by itself. The **a1** and α2 proteins belong to the homeobox family, and the homeoboxes fold the same way as described earlier. The α2 homeodomain, however, has a longer C-terminal tail that becomes a helix upon dimerization with the **a1** homeodomain (Figure 6.19).

The interaction within MATα2 as revealed at the 3-D level shows some interesting features that account for the specific interaction of these factors with **a**-specific operator DNA. An MCM1 dimer binds two α2 monomers (the cis and the trans). The amino terminal extension of the MATα2 homeodomain forms a beta-hairpin, which grips MCM1 with hydrogen bonding and close-packed side chains. The two proteins are also brought close to each other by MCM1-induced DNA bending; however, the complex has an unusual feature that can account for the interaction of the complex with the operator. A particular sequence (called the chameleon sequence) of the homeodomain adopts an alpha helical conformation in one monomer and a beta-strand conformation in the other. This change in conformation allows the complex to bind the **a**-specific operators in which the central 16-bp MCM1-binding sites are separated from the α2 binding sites by 3 bp on the one side and 2 bp on the other (Figure 6.19C).

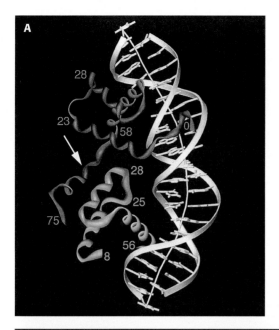

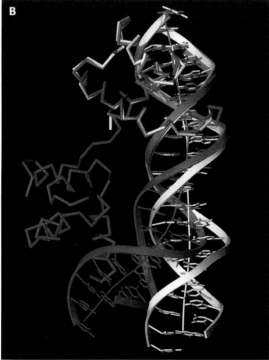

Figure 6.19. A: 3-D structure of the **a1**/α2-DNA ternary complex. The DNA is a haploid gene operator. α2 is red, and **a1** is blue. The numbers represent the positions of the corresponding amino acids in both homeodomains. The C-terminus tail, which upon dimerization has assumed a helical conformation, is shown by an arrow. C. Wolberger, Science 270: 262–9 (1995). **B:** Note that, when the heterodimer is bound to DNA (red), DNA is bent by 60 degrees when compared with the binding of α2 alone (gold complex). C. Wolberger, Science 270: 262–9 (1995).

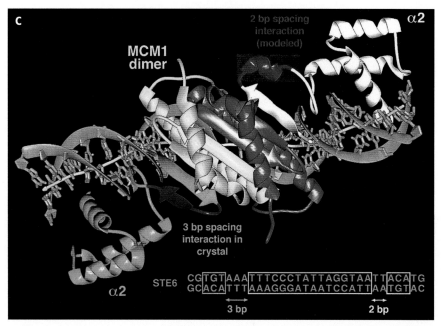

Figure 6.19 (cont.) C: A model for interaction of two α2 monomers with the MCM1 dimer and a 31-bp operator. Note that in the cis α2 monomer the chameleon sequence interacts with the 3-bp spacing as a beta strand, and that in the trans α2 monomer the chameleon sequence interacts with the 2-bp spacing as a helix. T. J. Richmond, Nature 391: 660–6 (1998). A and B reprinted with permission from American Association for the Advancement of Science. C reprinted by permission from Nature, Macmillan Magazines Ltd.

This example with the MAT products nicely illustrates how two homeobox-containing factors interact to regulate transcription and determine cell fate and differentiation. It is conceivable that different Hox gene products are engaged in similar interactions during the more complex events of development. Likewise, interactions among homeodomain-containing proteins might not be restricted to the homeodomain only.

STRUCTURAL FEATURES OF CLASS II DNA-BINDING DOMAINS

Previously we examined the role of some activators in specific class II gene regulation. It is obvious that, for such specificity, the activators must interact with the DNA in a unique way. This interaction is achieved by the characteristic DNA-binding domain of these factors and its mode of interaction with DNA. The two major groups of DNA-binding domains are the helix-turn-helix domain, especially the one found in homeoboxes, and the zinc-binding domains. Next we will examine the 3-D structure of these domains and their interaction with DNA. This study will help us understand how these factors can regulate genes in a specific manner. We will also examine some other DNA-binding motifs found in other activators.

The Helix-Turn-Helix Motif

In Chapter 5, we dealt with the 3-D structure of the bacterial repressors. We learned that the DNA-binding domain of the repressor is a helix-turn-helix motif. Also, we mentioned that the homeodomain binds to the DNA with a similar motif. The homeodomain is composed of 60 amino acids, which fold as three helices. Helices 2 and 3 create the helix-turn-helix motif, with helix 3 being the one that interacts with the major groove of DNA. The structural features of the homeodomain's HTH motif are very similar to its counterpart from the bacterial repressors. One of the differences is that the helix 3 of the homeodomain is longer. Another difference is that the operator is rather bent when compared with the DNA bound by the homeodomain (Figure 6.20).

As mentioned earlier, there are many Hox genes in vertebrates, whose product, the homeodomain, is very similar in sequence, structure, and DNA-binding specificity. How is such a high degree of biological specificity in gene regulation achieved? The answer could lie in the interactions among Hox products, especially those sequences outside the homeodomain. We have already seen how the interaction of the yeast MAT products can regulate cell differentiation. Likewise, in vertebrates, Hox genes can form heterodimers with a Hox cofactor, namely Pbx1 (in *Drosophila* the cofactor is the extradenticle). Pbx1 also contains a homeodomain, which is structurally slightly different from the other homeodomains. It contains a 3-amino acid (leucine, serine, asparagine) insertion at the C-terminus of helix 1. Additionally, it contains 13 residues that

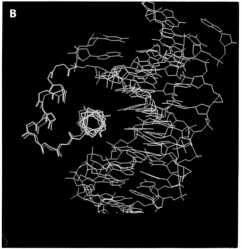

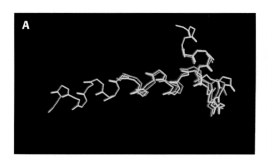

Figure 6.20. A: 3-D structure of the helix-turn-helix motif of the engrailed homeodomain (yellow) and its comparison with the same motif from the lambda repressor (orange). **B:** A view of repressor-operator (orange-blue) and homeodomain-DNA (yellow-purple) complexes. Note the difference in DNA bending, despite the similarity of the HTH motif. C. Pabo, Cell 63: 579–90 (1990). Reprinted with permission from Elsevier Science.

form a turn with a small helix and a fourth helix. In 3-D structural studies, it has been shown that HoxB1 forms an heterodimer with Pbx1 by the binding of an hexapeptide, found in the N-terminus of HoxB1, to a pocket of Pbx1 formed primarily by the three-amino-acid insertion (Figure 6.21).

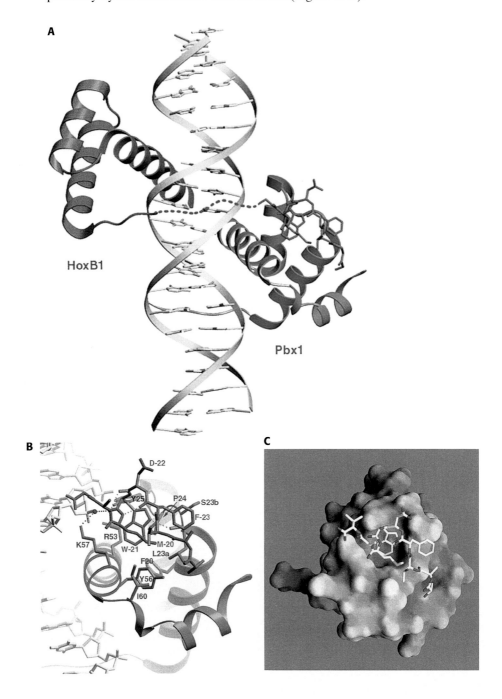

Even though there is little information at the 3-D level dealing with Hox-Hox interactions, we can gain some insights by carefully inspecting the sequence of the different Hox genes. As mentioned previously, vertebrate Hox genes are found in four clusters, with each cluster containing Hox 1 to 13 genes. Because the whole single cluster was duplicated, the paralog groups (i.e., A1, B1, C1, and D1 are paralogs) must be most conserved when their sequences are compared. The different paralog groups are characterized by differences in sequences either inside or outside the homeodomain. For example, paralog groups 1, 2, 3, 12, and 13 have little or nothing in common outside the homeodomain. Also, these groups can be distinguished by unique sequences, which are not implicated in DNA binding. Contrary to that, paralog groups 4 to 9 contain similar sequences outside the homeodomain. These differences suggest that unique sequences might be oriented away from DNA and involved in protein-protein interactions. Indeed, a mutation in HoxD13 (a gene involved in the morphogenesis of digits), located in the N-terminus and not the homeodomain, results in abnormalities of the fingers. This mutation does not affect the DNA binding of the protein but produces an alaline expansion, which might indicate that other protein-protein interactions are affected.

The HTH motif is not unique to homeodomains. Other proteins do contain this motif. Examples are the activator myb and the telomere-binding protein RAP1 (see Chapter 3). A similar motif, the helix-loop-helix (HLH) is found in other transcriptional activators. The difference is that the space between helix 2 and helix 3 is longer in HLH.

A similar, but distinct HTH motif is found in the interferon regulatory factor 1 (IRF-1). This factor is a member of the interferon regulatory factor family, which regulates interferon-responsive genes in response to infection by viruses. The 3-D structure of IRF-1 bound to regulatory element of the interferon beta promoter revealed that the HTH motif latches into DNA through three out of five conserved tryptophans (Figure 6.22).

The Zinc-Binding Domains

The zinc-binding domains were introduced in this chapter as the major DNA-binding domain of TFIIIA. This zinc finger, as it is also called, is

←

Figure 6.21. **A:** The human HoxB1-Pbx1-DNA complex. Note the interaction between the N-terminus hexapeptide of HoxB1 and the pocket at the C-terminus of Pbx1 helix 1. Also, note the characteristic three-helix fold of the HoxB1 homeodomain and the 13-residue extension (turn and small helix) of the Pbx1 homeodomain. **B:** Details of the interactions between the HoxB1 hexapeptide (red) and Pbx1 (blue). In the pocket the main interactions are manifested by the HoxB1 tryprtophan (−21) with phenylalanine 20 of helix 1, leucine 23 of the insertion, proline 24, and tyrosine 25 following the insertion and with arginine 53, tyrosine 56, and lysine 57 of helix 3. **C:** The interaction as seen by a solvent-accessible surface model of Pbx1 with the HoxB1 hexapeptide. Note the pocket where the tryptophan is buried. C. Wolberger, Cell 96: 587–97 (1999). Reprinted with permission from Elsevier Science.

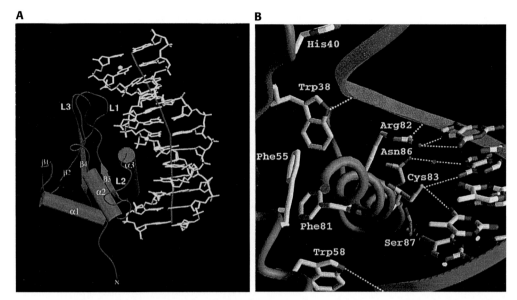

Figure 6.22. A: Interactions of IRF with DNA regulatory element. Note the embedding of helix 3 into the major groove of DNA. The DNA is bent 22 degrees. This bending might be different when other members of the IRF family interact with DNA. **B:** Details of the interaction. Note the interactions of helix 3 residues with the bases of the major groove and the backbone contacts of two tryptophans (Trp-38 and Trp-58; the interaction of Trp-11 is not shown). The tryptophans are fixed in that position by hydrophobic interactions with Phe-55 and Phe-81. A. Aggarwal, Nature 391: 103–6 (1998). Reprinted by permission from Nature, Macmillan Magazines Ltd.

characterized by the fact that the zinc ligands are two cysteines and two histidines. However, different zinc-binding domains are found in transcriptional factors in which the zinc ligands are four or even six cysteines. Because of this, the zinc-binding domains are majorly divided in three classes. Class 1 contains the C_2—H_2 domains, similar to the ones found in TFIIIA. Class 2 contains the C_2—C_2 domains called loop-zinc-helix or zinc twist. Class 2 domains are found in steroid receptors. Class 3 contains the zinc cluster where six cysteines are the ligands for two zinc atoms. Such a domain is found in GAL4.

The Class 1 Zinc-Binding Domains

Let us examine the 3-D structure of the zinc-binding domains from two different proteins. One of them is the mouse Zif-268 and the other is the human Gli protein. This comparison will help us identify the unique features in both proteins that account for their distinct function.

Zif-268 contains three zinc-binding domains, which bind a 10-mer palindromic sequence. The 3-D structure of these fingers bound to DNA shows that the three fingers wrap around the palindromic sequence (Figure 6.23).

A detailed examination of the contacts between the residues and the DNA reveals some very interesting patterns. It seems that there is some kind of

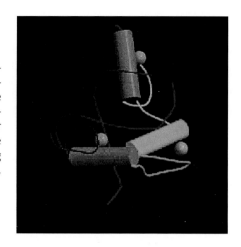

Figure 6.23. Interactions of the three zinc fingers of Zif-268 with duplex DNA. The three fingers are colored blue, green, and orange. The zinc is gray, and the DNA is red. Note the characteristic structure of the class 1 zinc finger with the two antiparallel beta strands and the following alpha helix that is the DNA-binding portion of the finger. M. Elrod-Erickson et al., PDB file 1A1F, Structure 6: 451–64 (1998).

periodicity in the way that particular amino acids from each finger interact with the DNA (Figure 6.24). In particular, the arginines in all fingers contact either the bases or the phosphates. The same is true for all aspartic acids that are found in the recognition helix. Another interesting feature is that all base contacts are only in one strand of the duplex DNA.

Gli is another protein that contains five zinc fingers. Gli is an oncogene and was discovered because it was amplified in human tumors. Gli also seems to play roles in the development of long bones. The 3-D structure of the five fingers

A ERPYACPVESCDRRFSRSDELTRHIRIHTGQK   Finger 1

    PFQCRI - - CMRNFSRSDHLTTHIRIHTGEK   Finger 2

    PFACDI - - CGRKFARSDERKRHTKIHLRQK   Finger 3

Figure 6.24. A: Residues from the three fingers of Zif-268 that are making contacts with the DNA. The two cysteines and the two histidines that are the zinc ligands are in bold, residues that make phosphate contacts are red, and residues that make base contacts are blue. **B:** The interaction of the three fingers with the DNA. Note that only bases of one DNA strand are contacted. Also note the periodic nature of the contacts in the protein residues and in the DNA. C. Pabo, Science 252: 809–17 (1991). Reprinted with permission from American Association for the Advancement of Science.

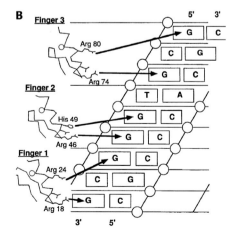

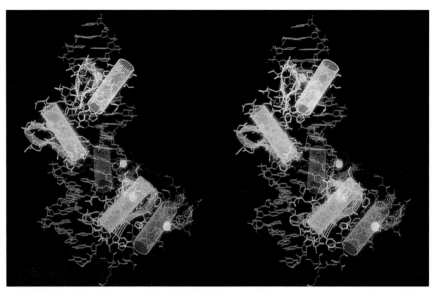

Figure 6.25. Stereo view of the Gli complex. Finger 1 is purple, 2 is green, 3 is red, 4 is blue, and 5 is yellow. DNA is dark blue. C. Pabo, Science 261: 1701–7 (1993). Reprinted with permission from American Association for the Advancement of Science.

complexed with DNA shows that fingers 2 to 5 fit in the major groove and wrap around the DNA for a full helical turn. Finger 1 does not make any contact with the DNA, but it does interact with finger 2. The overall arrangement is similar to that seen in Zif-268. The fingers bind in the major groove with the N-terminus of their alpha helices closest to the bases. The most extensive contacts are made by fingers 4 and 5 (Figure 6.25).

Let us now compare the zinc finger-DNA interactions in Zif-268 and in Gli. For this, in Figure 6.26, the amino acid sequence from the five Gli fingers are presented, and the contacts with bases and phosphates are marked as in Figure 6.24. We can notice some interesting similarities. For example, four residues,

---

Figure 6.26. A: Alignment of the three Gli fingers (3, 4, and 5) showing the residues that make contact with the DNA. The color code of the type of contact is the same as in Figure 6.24. Note that residues in similar positions in Zif-268 and Gli interact with the DNA. **B:** Summary of phosphate and base contacts by residues of the five Gli fingers. In comparison with A, note that similarly positioned amino acids contact same bases. **C:** Comparison of docking arrangements in Zif-268 and Gli complexes. The subsites are colored dark blue and have been superimposed. The three Zif-268 fingers are purple, Gli finger 2 is green, finger 3 is red, finger 4 is light blue, and finger 5 is yellow. Only the second beta strand and the helix are shown. Note that even though the three Zif-268 fingers (purple) are docked the same way, when compared with the Gli fingers there are some differences. Most of the differences can be seen with fingers 4 and 5, which are farther away from their subsites. C. Pabo, Science 261: 1701–7 (1993). B and C reprinted with permission from American Association for the Advancement of Science.

A PHKCT FEGCRKSYSRLENLKT HLRS -HTGEK

PYMCEHEGCSKAFSNASDRAKHQNRTHSNEK

PYVCKLPGCTKRYTDPS SLRKHVKTVHGPDA

B
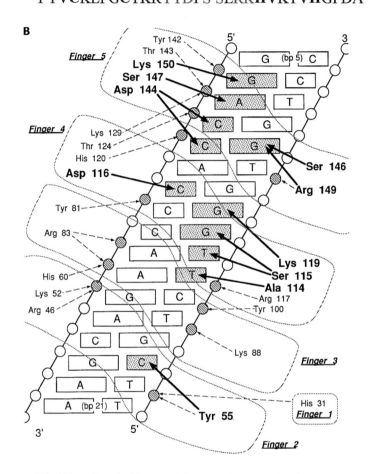

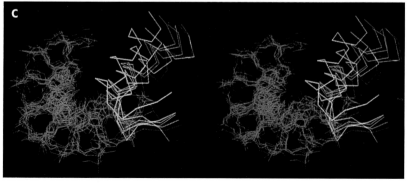

positioned similarly on the alpha helix, provide the critical contacts. There is also a correlation between the position of the residue on the alpha helix and the position of the base (within the finger's subsite; that is, the sequence that each finger spans in interacting) that it contacts. This is more spectacular for Zif-268 where the similarly positioned arginines contact a G in each subsite (Figure 6.24B). In Gli, we can see that similarly positioned lysines and serines in fingers 4 and 5 contact a G (Figure 6.26B). However, there are differences as well. The majority of Zif-268 contacts involve only one strand; however, Gli contacts involve both strands. There are interactions from Gli residues that are not seen in Zif-268. Also, docking of the Zif-268 and Gli fingers on their subsites show similar arrangements, but differences also surface when we compare individual fingers (Figure 6.26C). For example, even though all three Zif-268 fingers dock very similarly on their subsites, when they are compared with the five Cli fingers, we can observe differences in translation and rotation (Figure 6.26C).

The comparison between Zif-268 and Gli zinc fingers suggests that a general code for zinc finger-DNA interactions is unlikely. However, the differences are attributes of uniqueness that characterize the interaction of different zinc fingers with DNA and, therefore, their specificity in gene regulation.

The Class 2 Zinc-Binding Domains

The zinc ligands in this class are four cysteines, and usually these domains are found in the DNA-binding region of steroid receptors. The DNA-binding region contains two zinc-binding domains. These two domains are not always the same (as far as the position of the cysteines is concerned); as a result, their 3-D structure is different. In general, these motifs assume the loop-zinc-helix or zinc twist conformation. Let us elaborate more on the 3-D structure of these motifs. In the estrogen receptor, for example, there is a difference in the spacing between the first and the second cysteine. In motif 1 (the first finger) there are two residues between the two cysteines, and in motif 2 (the second finger) there are five. Likewise, this change would result in a slightly different 3-D conformation in motif 1 and motif 2. An important difference between class 2 and class 1 zinc fingers is that in class 2 the recognition helix spans residues found in the C-terminus of the finger in relation to class 1. As a result, the zinc is located at the beginning of the helix, while in class 1 it is located at the C-terminus of the helix (Figure 6.27).

The 3-D structure of the DNA-binding region of the steroid receptors shows that the two helices cross each other at midpoint and that the zinc holds the base of the loop at the N-terminus of each helix (Figure 6.28). The steroid receptors bind to DNA as dimers. The two monomers interact in such a way that their recognition helices are antiparallel and at such an angle to the DNA axis that they bind symmetrically to successive major grooves located 34 Å apart. This is a perfect arrangement for binding palindromic sequences that characterizes the HRE (see previous sections). Mutagenesis studies have shown that there

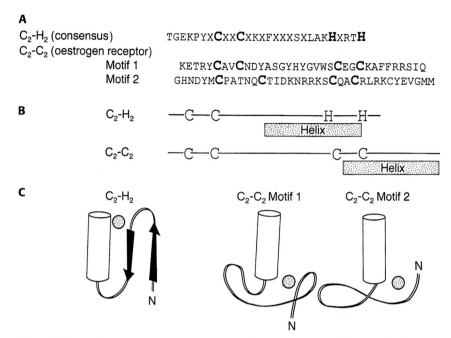

A

C$_2$-H$_2$ (consensus)
C$_2$-C$_2$ (oestrogen receptor)

TGEKPYX**C**XX**C**XKXFXXXSXLAK**H**XRT**H**

Motif 1
Motif 2

KETRY**C**AV**C**NDYASGYHYGVWS**C**EG**C**KAFFRRSIQ
GHNDYM**C**PATNQ**C**TIDKNRRKS**C**QA**C**RLRKCYEVGMM

B

C$_2$-H$_2$

C$_2$-C$_2$

—C—C————————————H——H—
Helix

—C—C————————————C—C—
Helix

C

C$_2$-H$_2$ C$_2$-C$_2$ Motif 1   C$_2$-C$_2$ Motif 2

Figure 6.27. A: Sequence and structural characteristics in class 1 and class 2 zinc fingers. On the top, the consensus sequence of class 1 and class 2 (human estrogen receptor) zinc fingers is presented, indicating that the ligands are two cysteines and two histidines (C$_2$—H$_2$) for class 1 and C$_2$—C$_2$ for class 2. **B:** The location of the recognition helix is shown for both classes. Note that the recognition helix for class 2 zinc fingers is located in the C-terminus of the finger. **C:** The difference in the 3-D structure between class 1 and class 2 zinc fingers. Note that in class 2 the antiparallel beta strands do not form, but instead we have a loop or twist conformation. Also note that the structure between motif 1 and motif 2 of class 2 zinc fingers is also different because of the different arrangement of cysteines in the primary sequence. D. Rhodes, TIBS 16: 291–6 (1991) and Nature 348: 458–61 (1990). Reprinted by permission from Nature, Macmillan Magazines Ltd. and Elsevier Science.

are, in fact, two regions of the receptor that contribute to the DNA binding and the discrimination of the different HRE. As discussed previously, the HREs for different steroid receptors are similar, but yet different. The difference can be found either in sequence or in the spacing between the half sites of the palindrome. In the estrogen receptor, the first region is part of the recognition helix of the first finger and of the residues glutamate, glycine, and alanine. For the glucocorticoid receptor, the residues are glycine, serine, and valine. The second region in the estrogen receptor is 5 amino acids long and is located in the spacing between the first two cysteines of the second finger. These regions are involved in the discrimination of half-site spacing (for example, estrogen receptor versus thyroid receptor, whose half sites have 3 and 0 spacing, respectively). These five amino acids are involved in protein-protein interactions that lead to dimerization (Figure 6.28). This activity suggests that interactions in

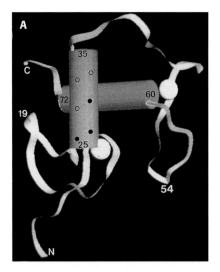

Figure 6.28. A: 3-D structure of the human estrogen receptor DNA-binding domain consisting of the two zinc fingers. Motif 1 is from 1 to 36 and motif 2 from 36 to 84. Note the crossing between the two helices (red), the zinc atoms (spheres) at the N-terminus of each helix, and the difference in the loop structure (green) in the two fingers. The dark dots indicate amino acids glutamate-25, glycine-26, and alanine-29, which are involved in HRE recognition. The open dots represent two conserved lysines and one arginine. **B:** A model of the receptor dimer. Note the relative orientation of the two monomers resulting in antiparallel recognition helices that can bind in successive major grooves. Conserved lysines and arginines are indicated by the pluses (+ + +). The regions that are involved in dimerization are bold, and zinc atoms are represented as spheres. D. Rhodes, TIBS 16: 291–6 (1991) and Nature 348: 458–61 (1990). Reprinted by permission from Nature, Macmillan Magazines Ltd. and Elsevier Science.

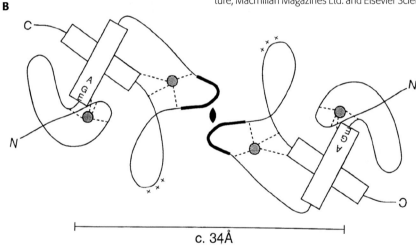

c. 34Å

this region influence the spacing and relative orientation of the two monomers and, therefore, the recognition of the half sites.

The Class 3 Zinc-Binding Domains

The main feature of these domains is the zinc cluster. Two zinc atoms are bound by six cysteines, thus creating a cluster. Such a zinc cluster is found in GAL4. The arrangement of the cysteines follows:

10-**A**C**D**I**C**RLKKLK**C**SKEKPK**C**AK**C**LKNWE**C**RY-40

This sequence creates a helix-loop-helix domain with the two zinc atoms located between the helices (at their N-terminus), and with helix 1 embedded in the

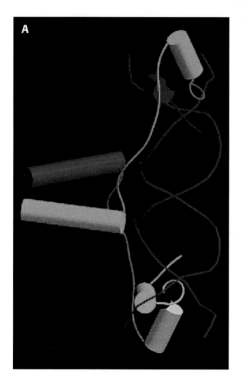
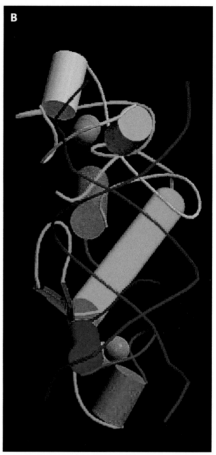

Figure 6.29. A: Binding of a GAL4 dimer onto DNA. The recognition helix is blue in one monomer and green in the other. The two zinc atoms in each monomer are represented as black spheres. Note the dimerizing alpha helices, which form a parallel coiled coil. R. Marmorstein et al., PDB file 1D66, Nature 356: 408–14 (1992). **B:** Binding of Ppr1 dimer onto DNA. The light green and dark blue helices are the recognition helices. Note that the 3-D structure is similar to that of GAL4. One major difference is that the dimer binds shorter DNA sequences mainly because, in contrast to GAL4, the extension between helices 2 and 3 is a loop. The zinc atoms are gray spheres. R. Marmorstein et al., PDB file 1PYI, Genes Dev. 8: 2504–12 (1994).

major groove of the UASs. The charged amino acids, especially lysines, seem to be important in DNA recognition. The two monomers dimerize with a third helix that is found next to the zinc cluster domain (Figure 6.29A). These helices form a parallel coiled coil. A similar zinc cluster is found in the pyrimidine pathway regulator (Ppr1) (Figure 6.29B).

Other DNA-Binding Domains

The helix-turn-helix motif and the zinc-binding domains can be found in many transcriptional factors and have been regarded as the main DNA-binding domains. However, many other transcriptional factors contain different DNA-binding domains that are unique to them as individuals or groups. The way that they bind DNA also differs according to their specificity. In the next few pages, we will examine some of these domains to provide a more in-depth view of gene regulation at the 3-D level. We will, however, limit these structures to a small number because exhaustive analysis would be overwhelming.

The Leucine Zipper

In proteins, amino acids with common chemical properties tend to segregate on the surface of alpha helices. This phenomenon is called amphipathy. An example would be the segregation of hydrophobic on hydrophilic amino acids. Early studies dealing with the structure of a protein that binds CCAAT, the CAT/Enhancer-binding protein (C/EBR) revealed such an amphipathic phenomenon. A segment of this protein contains four leucines found every seven amino acids. The four leucines (hydrophobic amino acid) line up in a column and occupy every seventh position in the helix (Figure 6.30).

After the discovery of the leucine repeat in C/EBR, similar consensus sequences were also found in other proteins. Two of them were the oncogenes Jun and Fos. These are nuclear proteins involved in tumor formation and are members of the transcriptional factor family that bind to the AP-1 site of gene promoters. Jun and Fos can heterodimerize, and the complex is also called AP-1. However, the leucine region is not the DNA-binding domain but rather serves to zip two such domains together and create homodimers or heterodimers via this leucine zipper. The two alpha helices zip in parallel and create a coiled coil. The DNA-binding domain is, in fact, a basic region that is found before

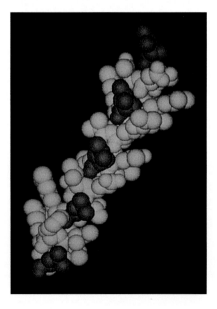

Figure 6.30. An amphipathic helix with the leucines (red) lined up. Courtesy of Dr. S. L. McKnight. S. L. McKnight, Scientific American 54–64 (April 1991).

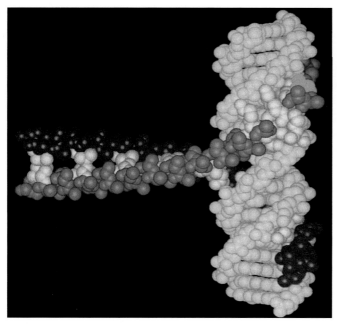

Figure 6.31. The 3-D structure of a zipper heterodimer. The one monomer is blue, and the other is red. The dimer is zipped by the leucines (yellow, left portion). The charged amino acids that bind to DNA are also yellow, and green is the invariant asparagine. Courtesy of Dr. S. L. McKnight. S. L. McKnight, Scientific American 54–64 (April 1991).

the leucine-rich region. This basic region contains lysines and arginines, which, being positively charged, can bind to acidic (negatively charged) DNA. Two groups of lysines or arginines in the basic region are, in fact, separated by an invariant asparagine, which breaks the helix and by bending allows it to bind to major grooves (Figure 6.31). The consensus sequence found in zipper proteins follows:

```
---------RR-RN-----R-R-RR------L------L------L------L------L
          KK K         K KK
```

The Immunoglobulin Fold in NFAT

As mentioned earlier, Jun and Fos heterodimerize and form the AP-1 complex that can bind specific sites in promoters. However, Jun-Fos can interact with other factors as well. Such a factor is the nuclear factor of activated T cells (NFAT). This factor is extremely important in the immune system. When an antigen is presented to the T-cell receptor, T cells are activated, and a cascade of transcriptional responses is induced. For example, synthesis of IL-2, which is an important early event, depends on NFAT binding to sites of the IL-2 promoter. Full response at NFAT sites, however, requires the concomitant activation of the AP-1 transcriptional factor family. In fact, there is a Jun-Fos site immediately downstream from most NFAT sites in the promoter region of IL-2 and other cytokine genes. The Jun-Fos heterodimer binds this site and, synergistically with NFAT, activates the expression of immune-response genes. This association of the three proteins is also achieved by bending Fos and DNA and creates a continuous groove for the recognition of 15 base pairs. This example not only will help us visualize a complex interaction for regulation but also will show how different activators work in unison to generate a transcriptional response. It illustrates another aspect in eukaryotic gene regulation.

NFAT is made up of two domains, each one having a structural homology with members of the Rel family. These domains are also called Rel homology regions (RHR). Rel proteins are structurally characterized by the immunoglobulin fold. One domain is the N-terminal domain, and the other is the C-terminal domain. Both domains are made up mainly of beta strands and loops, with two small helices that virtually join the two domains. In fact, it seems that, upon DNA binding, there are conformational changes in the RHR-N (mainly in the helices). Also for interaction with Fos and Jun both Fos and Jun must bend toward NFAT (Figure 6.32).

The N-terminal domain of NFAT binds DNA almost exclusively, and the binding is in the major groove. The binding involves residues of the AB loop

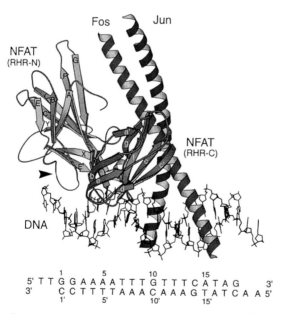

```
            1         5        10         15
      5' T T G G A A A A T T T G T T T C A T A G     3'
      3'     C C T T T T A A A C A A A G T A T C A A 5'
            1'        5'       10'        15'
```

Figure 6.32. The NFAT-AP-1-DNA complex. RHR-N is yellow with its beta strands marked with capital letters, and RHR-C is green with its beta strands marked with small letters. The NFAT site is GGAAAA, and the AP-1 site is TGTTTCA. The arrowhead indicates the loop between A and B strand that is involved in DNA binding. S. C. Harrison, Nature 392: 42–8 (1998). Reprinted by permission from Nature, Macmillan Magazines Ltd.

(arginines), the E′F loop, and the alpha1 and alpha2 helices. Only one residue of the C-terminal domain (a lysine of the fg loop) interacts with DNA. Interactions with Fos are with residues of the CX loop (of the RHR-N) and the cc′ loop (of the RHR-C). Interactions with Jun are exclusively with residues of the E′F loop, which also includes a part of helix alpha1 (Figure 6.33).

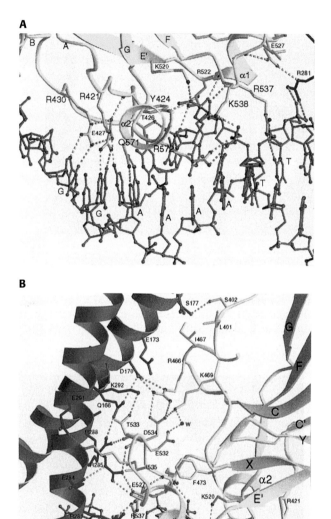

Figure 6.33. A: Interactions of NFAT with its DNA site. Note that residues in the AB and E′F loops and the helices mediate most of the interactions. **B:** Interactions of NFAT (yellow) with Jun (blue) and Fos (red). Note the interactions of NFAT with Jun through E′F loop residues and with Fos through CX loop residues. S. C. Harrison, Nature 392: 42–8 (1998). Reprinted by permission from Nature, Macmillan Magazines Ltd.

The T Domain

The T domain is a 180 amino-acid region with DNA-binding properties. This domain is found in the so-called T-box genes, a family whose prototype is the Brachury gene, which is of foremost importance in embryonic development. Some T-box genes also play important roles in differentiating arms from legs. The T domain is bound to palindromic DNA as a dimer and interacts with both major and minor grooves. A very unique feature of this domain is that a C-terminal helix is embedded into an enlarged minor groove, without bending the DNA. The residues of this helix interact by hydrophobic contacts with the sugar backbones and the edges of the bases. Specifically, two phenylalanines reach to the bottom of the minor groove and one of them (Phe-211) contacts a base by hydrogen bonding (Figure 6.34).

A careful observation of the T-domain structure shows that there is some structural similarity with the N-terminal domain of proteins sharing the immunoglobulin fold, such as NFAT. Interestingly, when the fold (seven-strand barrel) is superimposed, the ab loop coincides with the AB loop found in the Rel protein family. Recognition of the major groove by both AB and ab domains is mediated by arginines.

Signal Transducers and Activators of Transcription Proteins

Gene regulation depends largely on extracellular signals that are transduced to the nucleus. Many of these signals (such as cytokines, growth factors, and hormones) act through receptors at the surface of the cell and specific proteins in the cytoplasm. Single transducers and activators of transcription (STAT) proteins are involved in specific gene regulation in response to extracellular signals. A signal is initiated at the cell surface through the binding of a factor to a receptor. Upon binding, the receptor is autophosphorylated at a tyrosine residue. STAT protein is then recruited from the cytosol and recognizes the phosphotyrosine through the so-called SH2 domain. Following this recognition, STAT protein is phosphorylated on a C-terminus tyrosine. Such phosphorylation could be from the receptor or a receptor-associated kinase. The phosphorylated STAT protein can then dimerize and translocates to the nucleus where it binds specific promoters in target genes.

The 3-D structure of STAT3beta has been determined and has provided very important information about its binding to DNA. STAT3beta protein has an N-terminal domain consisting of four antiparallel helices connected by short loops. This domain is very elongated, with helices alpha1 and alpha2 spanning the entire domain. An eight-stranded beta barrel follows this four-helix domain. Strands a, b, and e form the one sheet of the barrel, and strands c, f, and g form the other sheet. The two sheets are connected by strands x/c′ through hydrogen bonding. The beta barrel is linked to the SH2 domain by a small helical domain (helices alpha5 to alpha8). The SH2 domain has a central three-stranded beta-pleated sheet (strands B, C, and D) flanked by helix alphaA and strands betaA and betaG (Figure 6.35A). The two monomers dimerize through their

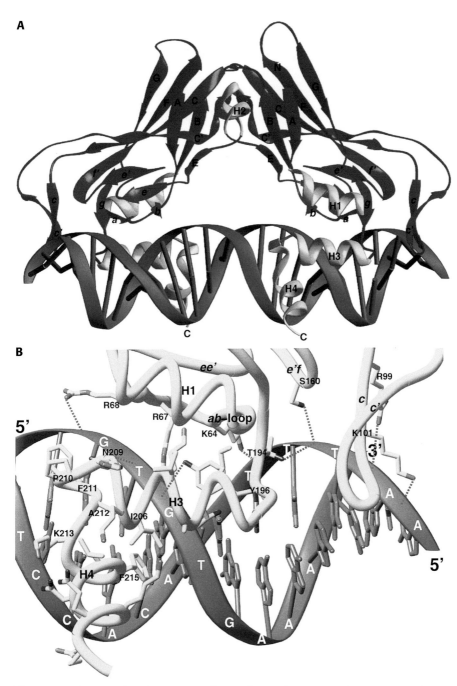

Figure 6.34. A: Interaction of a Xenopus T-domain dimer with DNA. Note that the C-terminal helix is buried in the minor groove. **B:** Details of the interaction of the C-terminal helix with the minor groove. Note the interaction of Pro-210, Lys-213, and Ala-214 with the sugar backbone and the edges of bases T, C, and C, respectively, and the interaction of Phe-211 and Phe-215 at the bottom of the minor groove. Phe-211 contacts, by hydrogen bonding, the G on the other strand. C. W. Muller, Nature 389: 884–8 (1997). Reprinted by permission from Nature, Macmillan Magazines Ltd.

A

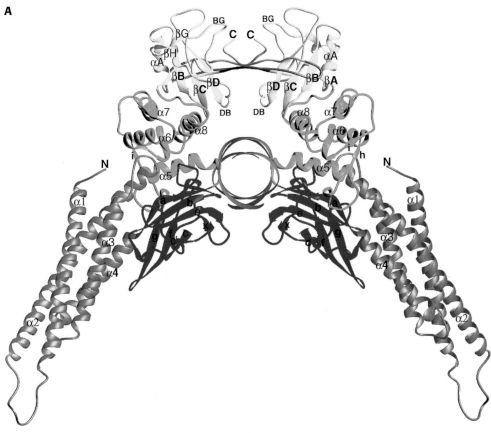

B

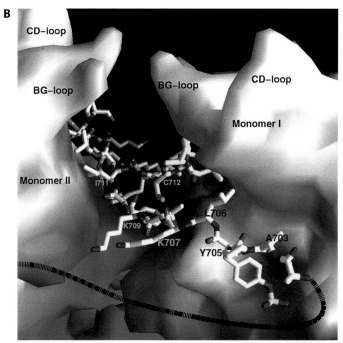

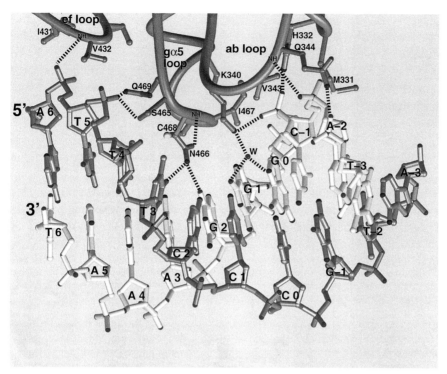

Figure 6.36. A 3-D model showing the interactions of ab, ef, and g/alpha5 loops from one monomer with DNA. The contacts are mainly with the phosphates. Note also the changes in the DNA. C. W. Muller, Nature 394: 145–51 (1998). Reprinted by permission from Nature, Macmillan Magazines Ltd.

SH2 domains. The phosphotyrosine of one monomer is bound to the SH2 domain of the other (Figure 6.35B).

DNA binding is mediated by four loops per monomer. Three of these loops (ab, cx, ef) protrude from the beta barrel domain, and one (g/alpha5) links the barrel with the connector domain. The majority of the DNA-binding residues are contributed by the ab and g/alpha5 loops, while loops cx and ef contribute one or two residues. The monomer sits above the minor groove. The critical residues seem to be Ser 465, Asn 466, Cys 468 and Gln 469 (all from g/alpha5 loop). All these residues form a hydrophobic binding pocket that accommodates the methyl groups of thymines in base pairs 3 and 4 (Figure 6.36). Also,

Figure 6.35. A: 3-D structure of mouse STAT3beta homodimer-DNA complex. The view looks down along the DNA axis. DNA is purple, the N-terminal 4-helix domain is blue, the beta barrel domain is red, the connector helical domain is green, and the SH2 domain and phosphotyrosine-containing domain is yellow. **B:** Covalently bound phosphotyrosine peptide by the SH2 domains. Carbon atoms linked to monomer I are yellow, and those linked to monomer II are white. Residues from alphaA (Lys), betaB (Ser), and BG loop (Ser) form direct polar interactions with phosphotyrosine (Tyr-705) and are strongly conserved in all SH2 domains. The hydrophobic residue leucine-706 (all STAT proteins have a hydrophobic residue at this location) packs against a hydrophobic pocket formed by residues in strands C and D. C. W. Muller, Nature 394: 145–51 (1998). Reprinted by permission from Nature, Macmillan Magazines Ltd.

Asn-466 seems to be important in determining base specificity. These residues also play roles in sequence-specific recognition by different STAT proteins. For example, Ser-465 and Cys-468 are replaced by valine and asparagine in STAT6, and Asn-466 (conserved in STAT1–4) is replaced with a histidine in STAT5 and STAT6. The bound DNA also deviates from its normal B-DNA-like conformation. It bends at the junction of the half sites, and the minor groove is larger and shallower (Figure 6.36).

Despite no obvious sequence homology, there is a striking structural and functional similarity between the STAT and the Rel family of transcriptional factors. The loops from the beta barrel that are used to contact DNA form a similar immunoglobulin fold. The STAT ab loop corresponds to the recognition loop in Rel proteins (such as NFAT), which we also saw in the T domain. Despite these similarities, there are also differences that account for different specificities. For example, NFAT interacts with the major groove, and STAT interacts with the minor groove.

MOST FREQUENT BASE-AMINO ACID INTERACTIONS

So far, we have seen that interactions of DNA and proteins are mostly mediated by electrostatic and hydrogen bonds between bases or phosphates and basic amino acids. However, other interactions are also important. These interactions are noncovalent and involve the electron clouds of the rings (pi-electron clouds). In the following table, we can see the frequency of base-amino acid interactions measured in 141 proteins, many of which are described in this book. As can be seen, the interactions between bases (especially purines) and arginine are the dominant ones.

Base-amino acid	Occurrence (%)
Gua-Arg	22
-Lys	7
-Asn	7
-Gln	4
Ade-Arg	17
-Asn	7
-Lys	6
-Gln	5
Thy-Arg	12
-Gln	4
-Lys	1
-Asn	1
Cyt-Arg	4
-Asn	2
-Gln	1
-Lys	0

CHAPTER SEVEN

Splicing

——— – – –

PRIMER Transcription produces an RNA transcript, which contains the entire nucleotide sequence as it is imprinted in the DNA. In other words, the primary RNA transcript also contains the intronic sequences. These intronic sequences must be removed to create a continuous coding sequence that will be able to be translated to a protein. The process of intron removal is called splicing. However, not all introns are spliced-out the same way. Depending on the mechanism involved in their splicing, the introns are divided into three groups. In this chapter, I will present the steps of intron removal, emphasizing the 3-D structure of RNAs and proteins involved. Some introns are removed with the help of proteins, but others self-catalyze their cleavage. Therefore, the 3-D structure of the introns must be important in splicing and will be presented in order to visualize the mechanism. The determination of the 3-D structure of self-catalyzed introns is another celebrated example of how the structure reveals the function.

Transcription results in the production of the primary RNA transcript that is a copy of the DNA sequences. These sequences also contain the introns that do not contribute to the translation process that decodes the DNA sequences and produce proteins. Therefore, the intronic sequences must be removed. This process of splicing is very complex and depends on the type of introns. Some introns require a complex machinery, composed of proteins and RNAs, called the spiceosome. These introns are contained in the nuclear RNAs, the precursors of all messenger RNAs (mRNAs). Other introns, the so-called group I and group II introns, are self-spliced and do not require a spliceosome. Group I introns are found in nuclear ribosomal RNA (rRNA), mitochondrial mRNA, and mitochondrial rRNA. Group II introns are found in mitochondrial mRNA. Before we elaborate on the differences between these introns and their mechanisms of splicing, let us consider some general features of the introns that account for splicing. Because the precursors of nuclear mRNAs contain many introns, their splicing should involve the same mechanism for all of them. It would be quite uneconomical if every intron had to be spliced employing a unique mechanism. The same should also hold true for group I and group II introns. After many sequences at the exon-intron border were available, it was obvious that the common splicing mechanism had to do something with consensus sequences found in the junction between the exons and the introns from both sides (the 5′ and the 3′).

The following table lists the consensus sequences in the different introns, and the exonic sequences are bolded.

Intron Type	5′ Splice Junction	Near 3′ Splice Junction	3′ Splice Junction
Nuclear pre-mRNA	**CRG**^GUAAGU	A	YnAG^**N**
Yeast	^GUAUGU	UACUAAC	YnAG^**N**
Group I	**U**^		**G**^
Group II	^GUGCG		YnAG^

Plastid genomes from *Euglena* and *Astasia* contain a unique class of introns called group III. These introns are short (91 to 119 nucleotides) and are rich in uridines. Their 5′ splice site is 5′-NUNNG, which is related to group II and the nuclear pre-mRNA 5′ splice sites. This similarity renders group III introns evolutionarily related to group II introns. The 5′ region of a group III intron might assume a stable stem-loop structure, which can provide flexibility in base-pair interactions with the 5′ exon, in much the same way that upstream exons interact with U5 snRNA (see later). This relation might indicate that group III introns could be close relatives to nuclear pre-mRNA introns that require *trans* acting RNAs. Also, other introns have been found to be interrupted by introns, and they are called twintrons.

Obviously, there are similarities and differences between the sequences found in the different intron types, and these differences have to do with the mechanisms involved in splicing as we will see later. The general idea for the splicing process is that the intron will be cut two times – first in the 5′ splice junction and second in the 3′. After removal of the intron, the two adjacent exons are joined together. Such removal of the introns is achieved by two transesterification events, which, in fact, are nothing more than a nucleophilic attack of a proton on the phosphodiester bond. We will elaborate on these events as we examine splicing in the different intron types in detail.

SPLICING OF THE NUCLEAR PRE-mRNA INTRONS

The general characteristic of this type of splicing is that the first transesterification event is mediated by an adenine, which is located near the 3′ splice junction. The OH group of this adenine attacks the phosphodiester bond at the 5′ junction between the exon and the intron, leaving the last nucleotide (usually a G) of the exon with an OH at the 3′ end. Next, the OH of the 5′ exon attacks the other end of the intron at the 3′ splice junction to cut the intron completely. Following the cleavage, the end result is the joining of the exons. The 5′ end of the intron is attached (via a 5′-2′ bonding) to the adenine, thus forming a lariat structure. For these transesterification events to occur, the ends of the two exons to be joined should come into the vicinity of each other. This is accomplished by the use of the spliceosome.

The First Transesterification

As already noted, the 2′ OH group of an adenine near the 3′ splice junction is responsible for attacking the 5′ exon-intron junction. This leaves the G (last nucleotide of the 5′ exon) with a free 3′ OH.

The Second Transesterification

The free 3′ OH group of the 5′ exon attacks the 3′ splice junction. The intron is completely spliced out, and the two exons are joined using the phosphate of the 3′ splice site.

Before we elaborate on the molecular mechanisms involved in this type of splicing, let us introduce the spliceosome. For the nucleophilic attacks to occur, the attacking and attacked molecules should, obviously, be close to each other. This is made possible by the spliceosome, a complex structure constituted of RNAs and several proteins.

The Spliceosome

We have two types of nuclear pre-mRNA introns. The major type (which is more common) has the sequence GT—AG found as consensus in the 5′ and 3′, respectively. Introns of the minor type have the sequence AT—AC at their ends. The major RNA components of the spliceosome are the small nuclear RNAs or U RNAs. We have U1, U2, U4/U6, and U5 for the major intron type and U11, U12, U4/U6, and U5 for the minor type. The spliceosome consists of these RNAs plus proteins, and these ribonucleoprotein complexes are called small nuclear ribonucleoptoteins (snRNPs). The snRNPs are called after their RNA components. For example, U1 RNP and U2 RNP. The protein component of snRNPs is classified into two groups. In group 1 we have specific proteins to each group, and in group 2 we have the core proteins that are common to all snRNPs. In Hela cells, seven such core proteins have been identified: B/B′, D1, D2, D4, E, F, and G. These proteins are also called Sm proteins because of their reactivity with the Sm serotype from patients with systemic lupus erythematosus. The core Sm proteins are assembled at the Sm site, a sequence rich in uridines, which is present in U1, U2, U4, and U5 snRNAs. Studies on the three-dimensional structure of two Sm complexes, the D3B and D1D2, have revealed how the seven core proteins might assemble around the Sm site. The heptamer has a ring structure that most likely surrounds the sequence of the Sm site (Figure 7.1).

Structure and Assembly of snRNPs

Even though the 3-D structure of snRNPs has not been solved as a whole complex, we do have some very informative data on their structure and assembly from electron microscopy. The U1 snRNP has three specific proteins – 70K, A, and C, where C is the smallest. U1 snRNP has a round main body and two small protuberances. A similar globular body can be seen in U4/U6

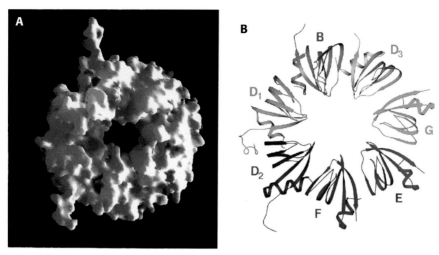

Figure 7.1. Higher order assembly of the human core snRNP proteins. **A:** Surface representation with electrostatic potential (blue, positive; red, negative). **B:** Ribbon model with the seven proteins designated with different colors. Note that the central hole is highly basic, indicating the site of interaction with U RNAs. K. Nagai, Cell 96: 375–87 (1999). Reprinted with permission from Elsevier Science.

snRNPs, which also has a Y-shaped domain protruding from the main body. The U2 snRNP when isolated under low-salt conditions has 12 specific proteins. At high-salt concentrations, most of the proteins are dissociated leaving only the A′ and B″ U2-specific proteins. U2 snRNP has two similar globular domains. U5 snRNP is the largest and contains nine specific proteins, some of which are very large. Its structure is elongated with a large head and a central body with a lower end. At low-salt concentration, U5 associates with the U4/U6 snRNPs with a set of five additional proteins. The U5.U4/U6 snRNPs has a triangular shape, and the lower part is similar to the lower part of U5 snRNP. The electron micrographs and schematic illustrations indicating the location of the specific proteins are shown in Figure 7.2.

Other Splicing Factors

Other proteins are also involved in splicing, even though they are not part of the snRNPs. These splicing factors have such different functions as RNA annealing activity, RNA recognition, pre-mRNA processing activity, and helicase activity. As will become apparent later, such activities are essential for snRNPs to function. The first class of proteins has RNA-annealing activity. These proteins contain RNA recognition motifs (RRMs), which bind single-stranded RNA, and RS domains (dipeptides of arginine/serine), which promote annealing of complementary RNAs. The second class contains members of a superfamily of ATPases. Some of these proteins contain the DEAD or DEAH sequence, and others contain the DECH or DECD sequence. Collectively, they are called

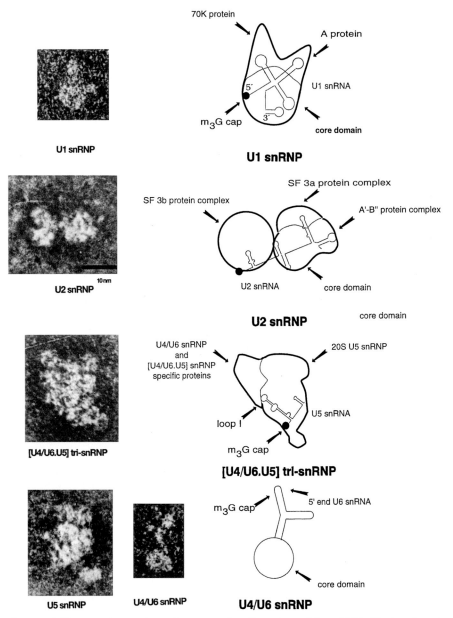

Figure 7.2. Electron micrographs and corresponding illustrations of the snRNPs. The location of the RNA component as well as the core and specific proteins is indicated. B. Kastner, in: RNP particles, splicing and autoimmune diseases (J. Schenkel, ed.) pp. 95–140, Fig. 5.4, Springer (1998). Reprinted with permission from Springer-Verlag.

the DExD/H-box proteins. This class contains many of the Prp (pre-mRNA processing) proteins and DNA helicases, as well as the initiation factor eIF-4A known for its RNA-unwinding activity. The third class includes one member of the GTPase superfamily that also shares extensive sequence similarity with EF-2, a protein required for the process of translocation during translation.

The Events of Splicing

Step 1. U1 snRNA contacts the 5′ splice site. Its 5′ sequences pair with sequences at the exon-intron border.

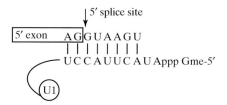

Step 2. U2 snRNA interacts with the branch site with an adenine (except in yeast where it pairs with UACUAC). In yeast, the branch site is first contacted by the branch-binding protein (BBP) and then by U2. Prp5 and 29 are needed for this interaction.

Step 3.1. U4 and U6 snRNAs form a complex with the help pf Prp24 (U4 + U6 → U4/U6).

```
5′         GAUCAGCA           GUUCCCCUGCAUAAGGAU        3′ U6
 3′        UAGGUCGU               AAGGGCACGUAUUCCU Appp Gme-5′ U4
```

Step 3.2. U5 snRNA interacts with U4/U6. For this interaction, several proteins such as Brr2 and Prp3, -4, -6, and -8 as well as ATP are needed.

U4/U6 + U5 → U4/46.U5

Step 4. Addition of the U4/U6.U5 complex into the spliceosome. When this happens, several RNA rearrangements must take place. First, the U6 snRNA interacts with the 5′ splice site, and U1 snRNA is displaced.

```
U6                                        5′
            GAGACA
    5′ exon  GUAAGU
```

Then, the U4/U6 interaction is interrupted, and U2 pairs with U6 to form the U2/U6 helix and also to form the catalytic site. Next, U5 pairs with the end of the 5' exon.

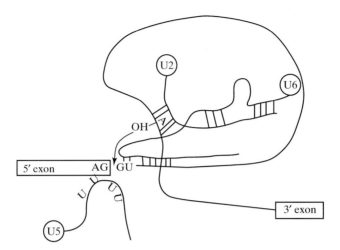

After all these interactions, the spliceosome is in place for the first transesterification, which also needs the presence of Prp2.

For the second transesterification, another rearrangement must take place with the help of Prp16, -17, -18, and -29. After the first transesterification, the 3' OH of the 5' exon is adjacent to the newly formed $2' \rightarrow 5'$ phosphodiester bond, which was the catalytic site. Just before the second transesterification event, the $2' \rightarrow 5'$ bond of the lariat intermediate must be replaced by the $3' \rightarrow 5'$ bond of the 3' splice site (Figure 7.3).

In a different model where we have two catalytic sites, the $2' \rightarrow 5'$ bond is fixed, but the 5' exon is repositioned near the 3' splice site. In both models, however, U5 snRNA interacts with sequences of the 3' exon (Figure 7.4). After the exons are joined, the spliceosome must be dissociated to its initial components. Prp22 and ATP are involved in this process. Furthermore, U2, U4, and U6 come back to their original configuration with the help of Prp43, and the lariat structure is denatured with the help of Prp26.

SPLICING OF GROUP II INTRONS

Group II introns are nearly 1,000 nucleotides long and are self-spliced, which means that a spliceosome is not necessary for splicing. In this sense, self-spliced introns can also act as enzymes, which is why they are also called ribozymes. Ribozymes owe their catalytic function to their specialized 3-D structure. Group II introns are structured with six domains. Domain 1 interacts with the 5' exon by base pairing. Domain 5 acts in *trans* to promote the hydrolytic cleavage at the 5' splice site. Domains 5 and 6 form the catalytic core in

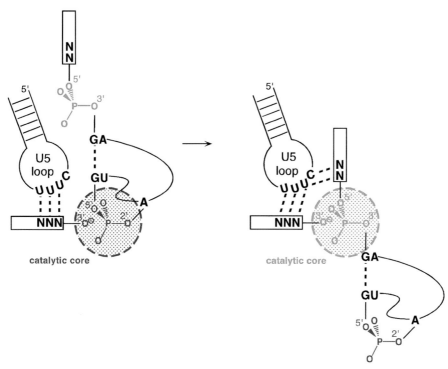

Figure 7.3. Replacement of the $2' \rightarrow 5'$ catalytic site before the second transesterification. C. Guthrie, Cell 92: 315–26 (1998). Reprinted with permission from Elsevier Science.

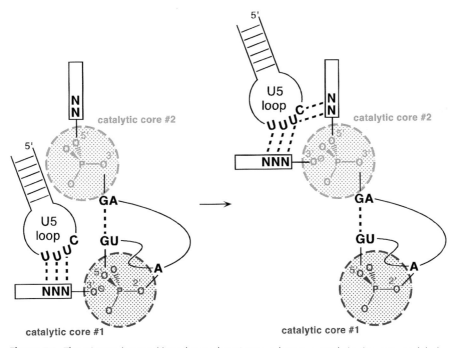

Figure 7.4. The 5' exon is repositioned near the 3' exon when two catalytic sites are modeled. C. Guthrie, Cell 92: 315–26 (1998). Reprinted with permission from Elsevier Science.

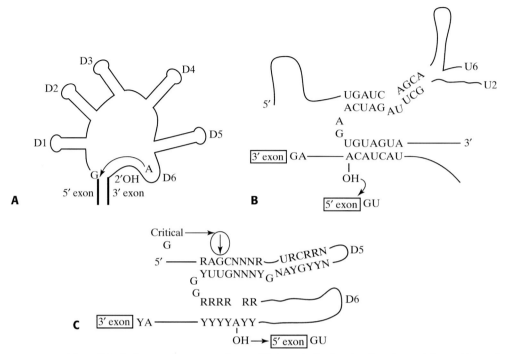

Figure 7.5. A: A general structure of group II introns indicating the domains and the positions of the 5′ and 3′ exons. **B:** Interactions of U2/U6 at the catalytic core of nuclear pre-mRNA introns. **C:** Interactions between domains 5 and 6 forming the catalytic core for the splicing of group II introns. Note the structural similarity between the two catalytic cores.

a similar structural fashion to the core made by the U2/U6 interaction during splicing of the nuclear pre-mRNA introns. In a manner similar to the splicing involving a spliceosome, a 2′-OH of an internal adenosine provides the attack on the 5′ splice site during the first transesterification (Figure 7.5).

Among group II introns from different species, the most conserved sequences are found in domain 5. Especially conserved is the sequence AGC (reminiscent of AGC found in stem 1 of the U2/U6 snRNA helix at the catalytic site of the spliceosome), with the G and its pairing to U being invariant. Indeed, this G seems to be the most important nucleotide for catalysis and interaction with the rest of the intron. Mutations in this G position do affect group II catalysis. To analyze the role of this G in the structure of the intron and in the catalysis, modified bases were used. Modified Gs were exchanged with the normal G, and the effects on ribozyme structure and function were assessed. Modifications included the N2, O6, or N7 substituents. It was found that N2 modification affected the minor groove, and, in turn, that affected binding of D5 to the core region, but not catalysis. On the other hand, modifications affecting O6 or N7 affected the major groove and resulted in loss of catalytic activity, but they had no effect on the binding (tertiary interactions) of the intron to

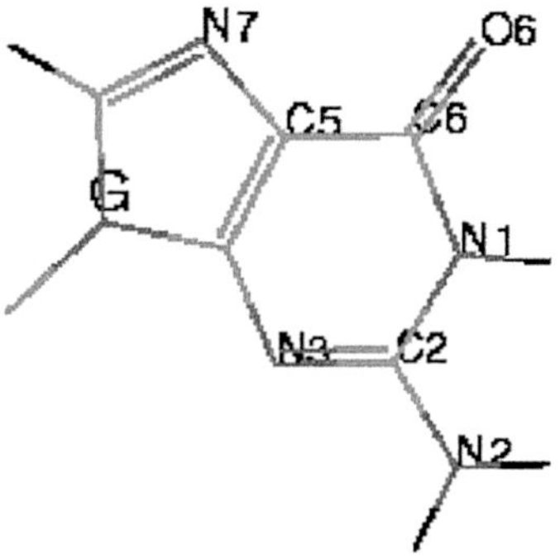

Figure 7.6. Modifications of the critical G in group II introns and the effects on the ribozyme structure and function. IMB-Jena.

	Minor Groove	Major Groove	Binding	Catalysis
N2	Affected	Not affected	Affected	Not affected
N7	Not affected	Affected	Not affected	Affected
O6	Not affected	Affected	Not affected	Affected

domain 5 (Figure 7.6). Considering data from other biochemical studies, it seems that major groove atoms participate in catalysis, despite their inaccessibility. One way to do this is by serving as sites for catalytic molecules or metal ions.

In the following 3-D structural models of D5, we can see the positions of important atoms that play roles in the aforementioned function of G and D5 (Figure 7.7).

SPLICING OF GROUP I INTRONS

Group I introns were discovered first in the 26S rRNA of *Tetrahymena*. In vitro these introns can be self-spliced; however, in vivo they may require proteins. They are nearly 400 nucleotides long (413 in *Tetrahymena*), and they are folded in a particular 3-D structure that creates the catalytic core. The folding is mediated by the formation of nine helical paired elements designated P1 to P9. The catalytic core is made out of P3, P4, P6, and P7. P1 contains the junction of the 5′ exon sequences and the beginning of the intron. The exonic sequence CUCUCU becomes part of P1 after pairing with the so-called internal guide sequence GGAGGG of the intron (Figure 7.8). In Figure 7.9, we can see the sequence of a part of the *Tetrahymena* 26S rRNA intron and how it folds in 3-D structure to create the catalytic core.

A major difference among group II introns is that the first transfer is mediated by a free guanosine and not an internal adenosine. The ribozyme, therefore, has two substrate-binding sites. One binds the free guanosine (G-site) and is located in the P7 and another binds the 5′ exonic sequences and is located in the P1. Group I introns are, in fact, metalloenzymes and require Mg^{2+} or Mn^{2+} for catalysis.

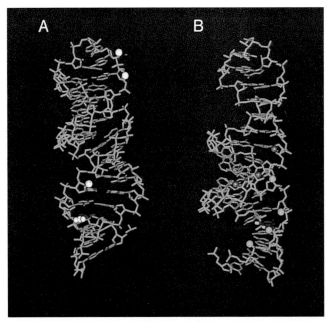

Figure 7.7. A: A 3-D model of D5 showing the critical G (red) with its N2 (three small white spheres) and its projection in the minor groove. Along with certain 2′-OH groups (larger spheres), they stabilize the binding of D5 to the rest of the intron by forming a continuous binding face. **B:** A 3-D model of D5 turned 180 degrees when compared with A to show the substituents in the major groove. G is red with O6 and N7 designated as yellow points. Phosphate oxygens (yellow spheres) and 2′-OH (orange spheres) known to affect efficiency of splicing define the chemical face of D5. A. M. Pyle, Mol. Cell 1: 433–41 (1998). Reprinted with permission from Elsevier Science.

The first step for autocatalysis is the binding of the free guanosine to the G-site C311-G264. A 3-D model of the interaction between P1 and the catalytic core reveals interesting structural and functional aspects of the autocatalysis. The core has an organized gateway to the G-site. The gateway is mainly created

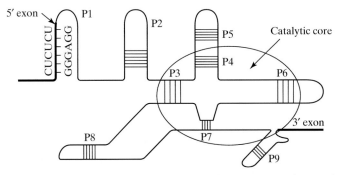

Figure 7.8. The common secondary structure of group I introns, showing the nine helical domains and the formation of the catalytic core.

A

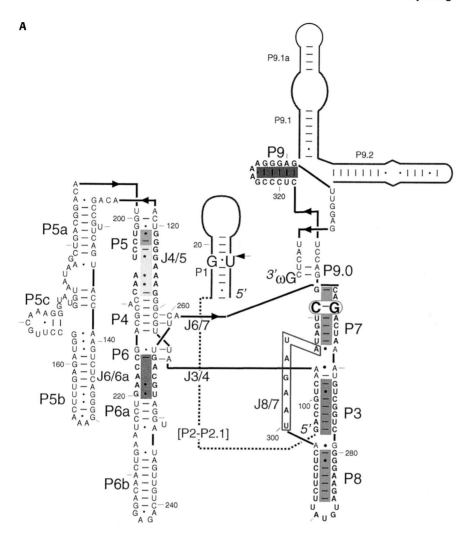

Figure 7.9. Secondary **(A)** and 3-D structure **(B)** of the *Tetrahymena* group I intron. P denotes the helical domains, and J identifies the joining regions. The 3-D structure does not include all the sequences presented in A. Note for example, in C, that P1 is not part of this structure. The ribozyme has a shallow cleft that forms the P1 binding site. Red represents direct contact, and green denotes the surface within 3A of P1. T. Cech, Science 282: 259–64 (1998). Reprinted with permission from American Association for the Advancement of Science.

by A306 of J8/7 (joining to P7), A261 of J6/7 (joining to P7), A207, and C208 of P5. It is believed that the phosphates of these four nucleotides form the ligands to the catalytic metals. Obviously, P1 and the 5′ splice site are placed very close to this guanosine gate (Figure 7.10).

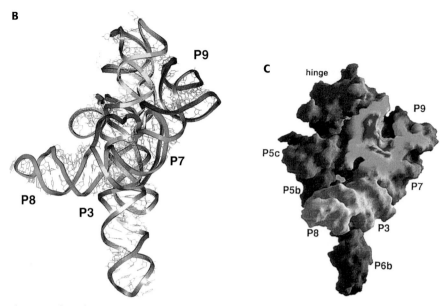

Figure 7.9 (cont.)

After the first transesterification, the free guanosine becomes the most 5′ part of the intron, and the 3′OH of the U from the 5′exon attacks the G414 that has now occupied the G-site. After this, the exons are joined, and the intron undergoes further processing that includes circularization and a third transfer (Figure 7.11).

The G-U base pair that defines the 5′splice junction in group I introns is also phylogenetically conserved. It is interesting to note that, in a manner similar to the G-U pairing found in group II introns, the N2 (exocyclic amide) of G22 presented in the minor groove contributes significantly to the binding by making tertiary interactions with the core of the ribozyme. In Figure 7.12, these minor groove determinants with which the core of the ribozyme interacts are shown on a 3-D model of the splice site helix.

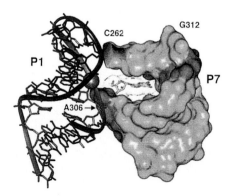

Figure 7.10. P1 and 5′ splice site interactions with the core of the ribozyme. P1 is close to P5 and J4/5. The G-site, C311-C264 is highlighted in darker green (as is the gateway nucleotide A306). The phosphate at the 5′ splice site is shown as a pink sphere. T. Cech, Science 282: 259–64 (1998). Reprinted with permission from American Association for the Advancement of Science.

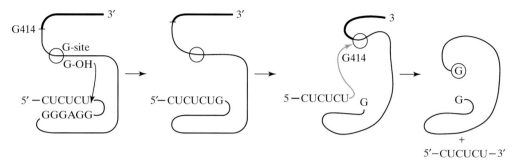

Figure 7.11. The events of splicing in group I introns.

OTHER RIBOZYMES

Except for group I and group II ribozymes, there are also others that have different structure, size, and mode of excision. Among them are the hammer-head, hepatitis delta virus (84 nucleotides), RNase P (350 to 410, the RNA subunit of eubacterial RNase P) and hairpin (50 nt, the minus strand satellite RNA of tobacco ringspot virus). For most of them (except RNase P), RNA cleavage produces a $2' \rightarrow 3'$ cyclic phosphate and a 5'-OH. The nucleophile is an adjacent 2'-OH. Structurally, the best characterized of these ribozymes is the hammerhead.

Structure of the Hammerhead Ribozyme

The hammerhead ribozyme is small (31 to 42 nucleotides); in fact, only 16 nucleotides are needed for its enzymic activity. It is found in viral satellite RNA in plants, newt satellite DNA, and viroids. The secondary and tertiary structures of the hammerhead ribozyme have shown that there are three stems with the catalytic site being a loop connecting to stem I (Figure 7.13).

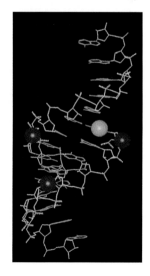

Figure 7.12. Groups of the P1 implicated in interactions with the core of the ribozyme. The G-U pair is shown in pink. Other 2'-OH groups (from G22, G25, and the exonic U3) are in red. T. Cech, Science 267: 675–9 (1995). Reprinted with permission from American Association for the Advancement of Science.

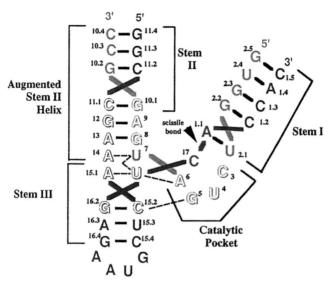

Figure 7.13. The structure of the hammerhead ribozyme, showing the catalytic site and the cleavage site between C17 and A1.1. W. G. Scott, Cell 92: 665–73 (1998). Reprinted with permission from Elsevier Science.

To understand the structural basis of hammerhead ribozyme self-cleavage, an experiment was designed where one of the residues involved in the cleavage (A1.1) was modified. The ribose of the leaving group was replaced with *tallo*-5′-C-methyl ribose (Figure 7.14A). This modification creates a ribozyme that has an unaltered attacking nucleophile but a modified leaving group that, in fact, results in inhibition of the cleavage. In other words, such modification can create an intermediate structure of the catalysis where possible snapshots of the action can be captured. Indeed, such an intermediate structure has been characterized at the 3-D level and provides a basis for self-cleavage of the hammerhead ribozyme. The intermediate structure differs from the ground state (uncleaved) by a conformational change observed in C17. C17, which provides the 2′-OH for the nucleophilic attack, swings out to stack upon A6, which also is stacked against G5 (Figure 7.14B). This conformational change produces a necessary geometry required for a successful nucleophilic attack and places the phosphate of A1.1 away from the A-form helical backbone phosphate geometry (nearly 90 degrees) and brings it closer to the geometry required for the attack (Figure 7.14C). These structural data, therefore, provide important evidence on conformational changes concerning the attacked phosphate that are necessary for self-cleavage. Metal ions are known to be important for self-cleavage of the hammerhead ribozyme. As can be seen in Figure 7.14B, a Co^{2+} ion is located near the 5′-oxygen of the leaving group. The position of this ion is near the scissile phosphate.

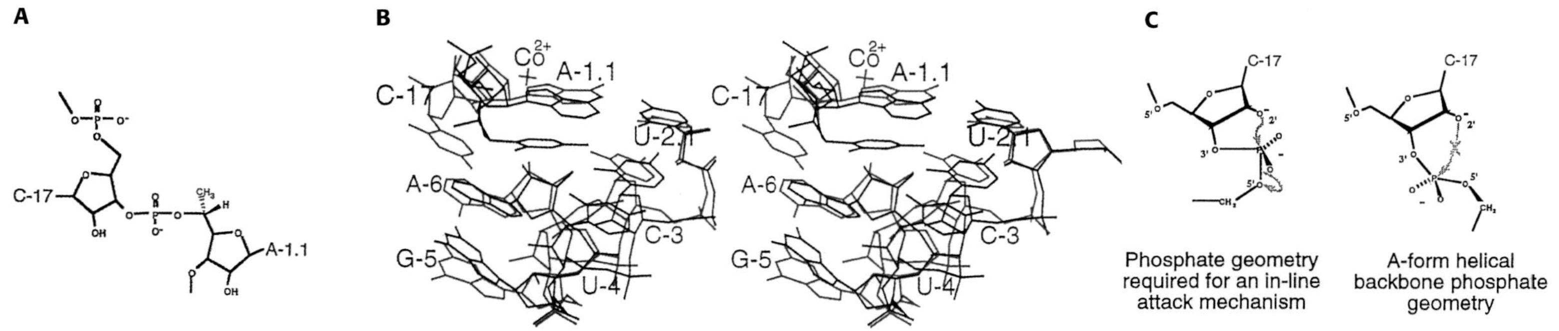

A

C-17

A-1.1

OH

OH

B

C-1.1   A-1.1
         Co²⁺

U-2

C-3

C-1.1

A-6

G-5

U-4

C-1.1   A-1.1
         Co²⁺

U-2

C-3

C-1.1

A-6

G-5

U-4

C

C-17

2'

5'

3'

O⁻

Phosphate geometry
required for an in-line
attack mechanism

C-17

2'

5'

3'

O⁻

A-form helical
backbone phosphate
geometry

Figure 7.14. A: The modification on the 5′ methyl ribose of the leaving group of A1.1. **B:** Conformational changes important for catalysis. The ground state structure (black) has been superimposed with the intermediate structure (red). Note the change in C17 that results in a geometry that is needed to bring the attacking nucleophile (2′-OH of C17) into the right position. **C:** On the left, we can see the correct position of the A1.1 phosphate of the intermediate structure. On the right, we see the standard A-form helical backbone phosphate geometry. W. G. Scott, Cell 92: 665–73 (1998). Reprinted with permission of Elsevier Science.

The Interesting Case of Group I Splicing by a Protein, Tyrosyl-tRNA Synthetase

As mentioned earlier, splicing of group I introns in vivo may also require proteins. Studies of mitochondrial RNA (mt RNA) splicing in *Neurospora crassa* and in yeast have pinpointed aminoacyl-tRNA synthetases (aaRSs) and other RNA-binding proteins that have adopted a function in group I splicing. Most likely these proteins recognize structures in the group I introns that resemble their RNA-binding sites. In the case of *Neurospora crassa*, it seems that tyrosyl-tRNA synthetase recognizes a tRNA-like structural motif in the two group I introns (LSU and ND1) in the catalytic core. These two *Neurospora crassa* group I introns have secondary and tertiary structures that are very similar to the *Tetrahymena* intron described earlier; LSU is also similar in size. Obviously, the catalytic core of these ribozymes is folded in much the same way as a tRNA. This will become apparent in Figure 7.15 where LSU and DN1

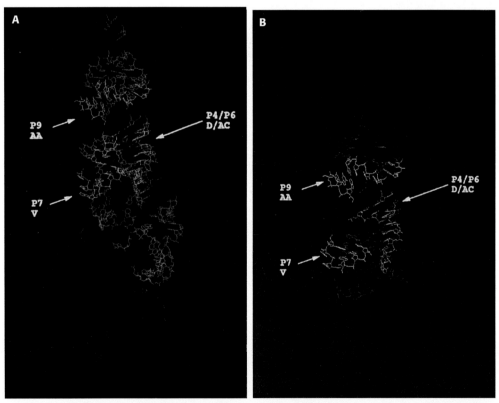

Figure 7.15. **A:** 3-D alignment of sites protected by tyrosyl-tRNA synthetase in the LSU intron (orange) and tRNA^Tyr (yellow). **B:** Alignment of protected sites in the ND1 intron (green) and tRNA^Tyr (yellow). Note that P9, P7, and P4/P6 are superimposed well with AA, V, and D/AC in both A and B. E. Westhof, Cell 87: 1135–45 (1996). Reprinted with permission from Elsevier Science.

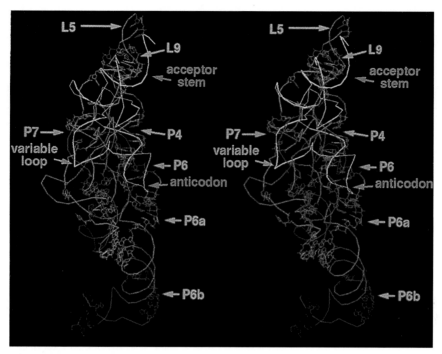

Figure 7.16. Stereo view of LSU and tRNA^{Tyr} superimposition. E. Westhof, Cell 87: 1135–45 (1996). Reprinted with permission from Elsevier Science.

structures are aligned with tRNA. In Figure 7.15, LSU or ND1 and tRNA^{Tyr} sites that are protected by binding with tyrosyl-tRNA synthetase are superimposed. It is evident that P9 aligns with the acceptor arm (AA), P7 aligns with the variable loop (V), and P4/P6 aligns with the anticodon arm (D/AC) (see also Chapter 10 for tRNA 3-D structure).

An alternative view of these alignments can be seen in Figure 7.16 in stereo. The phosphodiester backbone of the tRNA^{Tyr} model has been superimposed on the LSU model (orange). The nucleotides protected by aaRS are shown. tRNA is blue with the sites protected by the aaRS shown as yellow. Note the similarity in folding especially where protected nucleotides are placed.

CHAPTER EIGHT

Modifications of mRNA

--- --- ---

_____

PRIMER Except for the removal of its introns, the RNA transcript also undergoes other modifications. One of these modifications is the capping process that adds a 7-meG in the beginning, and another is the polyadenylation process that adds As at the end of each transcript destined to become mRNA. These modifications are mediated by the actions of specific enzymes. First, I present the mechanism of capping as revealed by the available 3-D structures of the enzymes involved. The same will be done for the addition of polyA, even though we have rather limited information about this modification at the 3-D level. Finally, other modifications such as RNA editing and RNA interference are mentioned.

THE CAPPING OF THE 5′ ENDS

Most of the eukaryotic mRNAs have their 5′ ends modified after transcription. This modification consists of an addition of 7-methyl guanylic acid linked to the transcript by a triphosphate group, the so-called cap. This reaction follows:

$$\text{Gppp}^{5'} + {}^{5'}\boldsymbol{ppp}\text{ApN} \rightarrow \text{meGp}\boldsymbol{pp}\text{Ap} + \text{pp} + \boldsymbol{p}$$

This guanylation reaction is catalyzed in three steps. First, an RNA 5′-triphosphatase removes the γ phosphate at the 5′ end of the transcript. Second, the enzyme GTP-RNA guanyltransferase adds a GMP residue to the 5′ diphosphate end in a 5′-to-5′ orientation. Last, an RNA (guanine-7-)methyltransferase adds the methyl group to the guanine. The molecular mechanisms underlying the actions of RNA 5′-triphosphatase and GTP-RNA guanyltransferase have

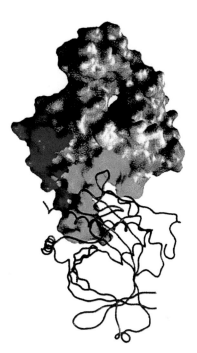

Figure 8.1. The three-dimensional structure of the yeast RNA 5′-triphosphatase. One monomer is shown as a surface representation (top), and the other is depicted as a worm trace of the polypeptide. The view is looking at the entrance of the tunnel. Green is the dimer interface, and red is the domain that binds guanyltransferase. S. Shaman, Cell 99: 533–43 (1999). Reprinted with permission from Elsevier Science.

been revealed at the 3-D level, and they have provided important clues on the mechanisms involved in the capping process.

Structure of the Yeast RNA 5′-Triphosphatase

The most prominent features of this enzyme are the presence of a tunnel where the mRNA transcript enters, an interface for dimerization, and a guanyltransferase-binding site. These structural features are depicted in Figure 8.1.

In the tunnel, we can observe interactions that are likely to play paramount roles in the cleavage of the γ phosphate of the transcript. A cross section of the tunnel shows the position of manganese, sulfate, and water. All these interact with residues of the protein, creating an octahedral geometry (Figure 8.2). The structure suggests that the γ phosphate occupies the same site as the sulfate ion, and the α and β phosphates are located in the entrance of the tunnel. In other words, the γ phosphate is placed deeper in the entrance of the tunnel. It also seems that the interaction with guanyltransferase creates a binding site for the RNA transcript that permits transit of the 5′ terminus from the triphosphatase site to guanyltransferase active site.

Structure of the Guanyltransferase

The enzyme can occur in two forms – open and closed. When the enzyme is open, the configuration is able to bind GTP. The enzyme has a U shape with two domains, 1 and 2. At this open state, the enzyme can bind GTP (b).

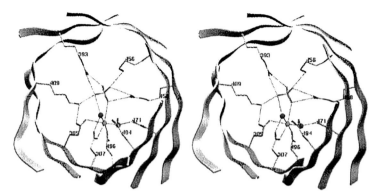

Figure 8.2. Stereo view of a cross section of the tunnel. Note the sulfate ion (stick model), manganese ion (blue sphere), and two water molecules (red spheres) that are necessary for catalysis. S. Shaman, Cell 99: 533–43 (1999). Reprinted with permission from Elsevier Science.

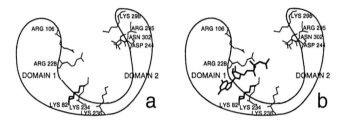

The binding of GTP stimulates closing, which is also stabilized by interactions between domains 1 and 2.

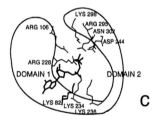

After closing, GTP is cleaved to produce GMP-adduct, which becomes attached to the enzyme by a lysine side chain.

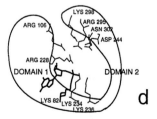

The cleavage subsequently destabilizes the interactions between domain 1 and 2, and the enzyme opens again.

After the enzyme opens, RNA can bind to GMP and create the complex between enzyme and capped mRNA. This sequence of events is also presented in Figure 8.3 with the 3-D models.

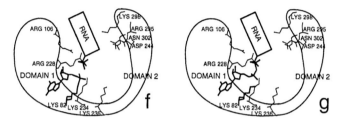

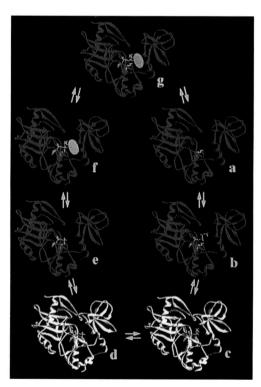

Figure 8.3. The sequence of guanylation by the enzyme at the 3-D level. Note that a through h are the same as in the in-text illustrations. D. B. Wigley, Cell 89: 545–53 (1997). Reprinted with permission from Elsevier Science.

In addition to capping, other modifications can occur with the addition of 2'-O-methyl groups on the first or the first two ribose residues of the transcript:

This type of methylation is catalyzed by a cap-specific nucleoside-2'-O-methyltransferase. This enzyme has the capability to recognize capped mRNAs, but it also needs to bind some of the sequence of the mRNA to anchor the target nucleotide to the active site. The 3-D structure of the vaccinia protein VP39, which acts as such an enzyme, shows this unique interaction with capped mRNA. The structure of the bound capped RNA is divided into three parts. The meG and the triphosphate bridge are recognized by two aromatic residues. The cap adopts an extended conformation. The meG is at the one end of the active site cleft, and the triphosphate is directed toward the active site (Figure 8.4). In the structure, the RNA has another six ribonucleotides. The first three constitute the second part of the bound structure. G1, A2, and A3 are stacked and have an A-form helical geometry. This trimer binds to the active site. The interaction with the protein is dominated by hydrogen bonding with the sugar-phosphate backbone only. The RNA backbone is turned between the third and fourth nucleotides. Thus, the second trimer (third part of the binding) is directed away from the active site (Figure 8.4).

Figure 8.4. **A:** The complex between VP39 and capped RNA. Note the catalytic site (coenzyme product S-adenosylhomocysteine; green) and the active site cleft where the meG and the first triplet are bound. Also note that the second triplet is directed away from the catalytic site. **B, C, D,** and **E:** Further details regarding the interactions between VP39 and capped RNA. In B, we can observe the interactions of protein residues and the triphosphate bridge. In C and D, we can see the stacked triplets and the interaction between the VP39 residues and the sugar-phosphate backbone. In E, we have the methyltransferase active site. **F:** Summary of all these interactions. F. Quiocho, Mol. Cell 1: 443–7 (1998). Reprinted with permission from Elsevier Science.

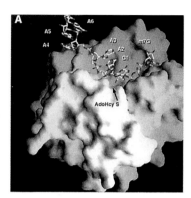

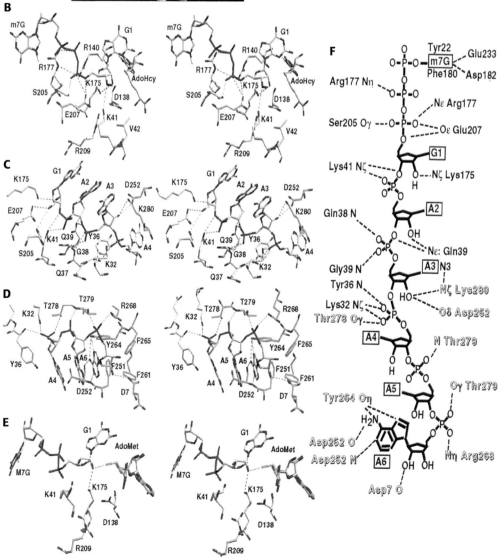

MATURATION AT THE 3′ END: ADDITION OF polyA

When the 3′ ends of genomic sequences are compared with those of cloned cDNAs, a clear difference emerges. The mRNA contains a tail of As that is not present in the genomic sequences. This tail instructs us that the polyA was added after the transcript was made. For this to occur, factors that recognize the right place at the 3′ and also appropriate enzymes for the synthesis of polyA must be in place. A careful examination of all 3′ genomic sequences indicates the presence of conserved regions that might serve as flags for factors to bind and initiate the process of polyA addition. Three such sequences have been implicated. The one is the AAUAAA, which is also called the polyadenylation signal. This sequence is always present upstream of the site where the mRNA is cleaved. The mRNA is usually cleaved at a CA, which is followed by a UGUGUG sequence. The AAUAAA is recognized by the cleavage polyA specificity factor (CPSF). Another factor, the cleavage stimulatory factor (CstF) recognized the UGUGUG sequence and also interacts with CPSF. Cleavage factor proteins (CFI and CFII) are also needed for the cleavage. After the transcript has been cleaved, a special polymerase, the polyA polymerase (PAP) adds the As at the end. Another protein, the polyA binding protein (PABII), helps to make the full-length polyA tail (nearly 200 nucleotides) (Figure 8.5). A different polyA-binding protein (PAB1) is important in interacting with eukaryotic translation initiation factors (see Chapter 10).

In addition to its function in transcription, RNA polymerase II also plays a very important role in the maturation of the 3′ end. It is known that an mRNA transcript made with pol I or pol III will never mature. This detail indicates that, somehow, RNA pol II plays a role in maturation. In fact, the C-terminal domain, which is unique to RNA pol II, has been found to be implicated as a cofactor

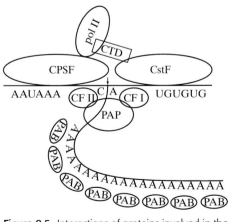

Figure 8.5. Interactions of proteins involved in the maturation of mRNA and polyA addition. See text for abbreviations.

for 3′ processing. CTD is associated with both CPSF and CstF (Figure 8.5). CTD can even facilitate cleavage in the absence of transcription. It is suggested, therefore, that the CTD is important for the assembly of the factors that are involved in the maturation of the 3′ ends.

3-D Structure of the polyA-Binding Protein

Comparison of sequences from many different polyA-binding protein (PABPs) has indicated the presence of four highly conserved RNA-binding domains or RNA recognition motifs. An RRM consists of nearly 100 amino acids with two conserved motifs, the RNP1 (Lys/Arg-Gly-Phe/Tyr-Gly/Ala-Phe-Val-X-Phe/Tyr) and the RNP2 (Leu/Ile-Phe/Tyr-Val/Ile-Gly-Lys-Asn/Gly-Leu/Met). It has been shown that not all RRMs are necessary for binding polyadenylated RNA. Experiments have shown that the first two RMMs support most of the biological functions of PABP. Four-stranded antiparallel beta sheet and two alpha helices characterize the 3-D structure of each RMM. The two RMMs are linked with a nine-residue linker. The structure creates a channel through which polyadenylated RNA passes. PolyA adopts an extended conformation in the channel, giving an antiparallel arrangement of protein and RNA. This means that the 5′ sequence binds RRM2 and the 3′ sequence binds RMM1 (Figure 8.6). The protein binds polyA through the conserved sequences in RNP1 and RNP2 with a combination of van der Waals contacts, hydrogen bonds, salt bridges, and stacking interactions. The concave surface of the protein is the polyA-binding surface. The convex surface contains a conserved region with hydrophobic/acidic sequence that might be involved in interactions with factors involved in the circularization of the mRNA (Figure 8.6).

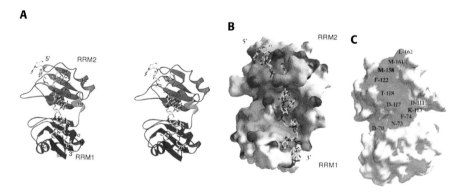

A **B** **C**

Figure 8.6. A: A stereo view of the human PABP RMM1/2-RNA complex. Note that the extension of the RNA is bound by the channel in the concave surface of the protein. **B:** Same complex as in A but shown as a solvent-accessible surface colored for the electrostatic potential. **C:** The convex site of the protein with a conserved region of hydrophibic/acidic residues colored green. S. Barley, Cell 98: 835–45 (1999). Reprinted with permission from Elsevier Science.

Maturation of mRNA Lacking polyA Tails

Many histone mRNAs are not polyadenylated. However, some histone mRNAs, whose expression is independent of the cell cycle control, do possess polyA. Also in amphibian oocytes histone, mRNAs have short polyA tails, but these are replaced later by mRNA lacking polyA. How then does maturation in these polyA-lacking mRNAs occur? A close inspection of the 3′ sequence of nonpolyadenylated histone mRNAs reveals that they possess the sequence CAAGAAGA, which in sea urchin, for example, is absolutely conserved. In higher vertebrates, the corresponding consensus sequence is 5′-A/GAAAGAGCUG-3′. The mature 3′ end is located upstream of this sequence. It is preceded by a sequence that allows the formation of a stem-loop structure. Even though this configuration might remind us of the Rho-independent termination structures we encountered in prokaryotes, cleavage to create the mature end must be achieved by other means. An snRNP that contains U7 is required. The 5′ sequence of U7 complements to the 3′ histone mRNA. Obviously, U7 forms a complex with the 3′ end of histone mRNAs, which is essential for cleavage (Figure 8.7). We do not know the exact mechanism, but an interesting speculation could be that cleavage is achieved by a transesterification event similar to that occurring in splicing.

As we know, RNA polymerase II also transcribes U RNAs. These U RNAs also lack polyA tails. Their 3′-OH ends are believed to be generated by transcription termination and by exonuclease action. Even though a sequence in U RNAs can form a stem and loop at the 3′ end, this structure does not seem to be essential for termination.

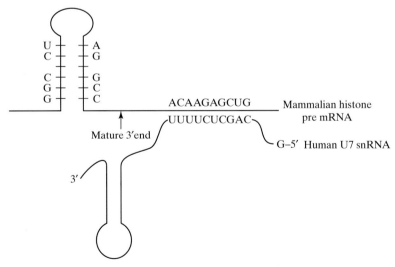

Figure 8.7. The structural features at the end of histone mRNAs and their possible interaction with U7.

RNA EDITING

A very interesting modification that changes the sequence of RNA transcripts is called RNA editing. The main manifestation of this event is either the change of one base to another (usually a replacement by U) or the addition of many Us. An example of a replacement of a base with a U is the editing of rat apolipoprotein-B mRNA. The unedited mRNA has a CAA triplet that codes for glutamine (residue 2153). This mRNA (encoding a 4563-residue protein) is normally expressed in the liver. In the intestine, however, the apolipoprotein-B mRNA encodes a 2153-residue protein. This encoding occurs because the CAA sequence has been changed to UAA, which creates a stop codon (Figure 8.8). Because there is no change in the DNA sequence for the apo-B gene, this type of editing is, in fact, a regulatory mechanism that allows different mRNAs to be specific for different tissues.

More extraordinary editing can be found in several genes of trypanosome mitochondria and paramyxoviruses. In trypanosomes, there are cases of addition of four uridines (coxII gene). The coxIII genes are even more spectacular in that more than half of the nucleotides in the mRNA consist of uridines that are not encoded in the DNA. How can the editing be achieved? Some possible ideas about the mRNA editing are presented in Figure 8.9. The main idea is that a specific RNA called guide RNA (gRNA) is complementary to the sequence that flanks the base (or the region) to be edited. Such gRNAs have been found, and they seem to be more homologous with the sequence downstream of the editing region than the upstream sequence. The gRNA is thought to form a duplex with its complementary mRNA. This would lead to cleavage of the mRNA by an endonuclease. Next, a terminal uridine transferase adds a U, followed by release of the mRNA, which now has been edited and has an extra U (Figure 8.9, left panel). Another model contains two cleavages and two ligations. According to this model, after the first cleavage, the U-containing 3′ end of gRNA ligates with the 5′ end of the restricted mRNA. A second cleavage adds one U to the mRNA, and ligation produces the edited mRNA (Figure 8.9, middle panel). A third model implicates two transesterification reactions and ligation steps. In this model, the cleavage of the mRNA is mediated by transesterification, when compared with the previous model (Figure 8.9, right panel).

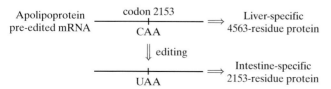

Figure 8.8. Editing in the rat apolipoprotein-B gene.

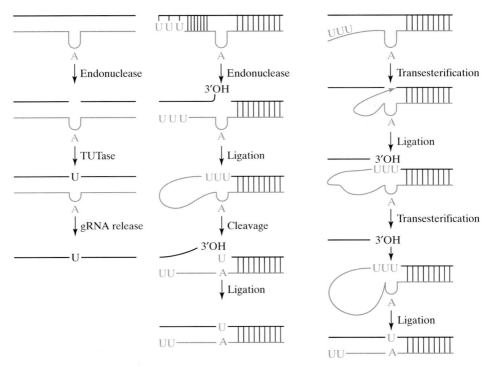

Figure 8.9. Models for RNA editing. Left panel: Addition of a U by the action of endonuclease and TUTase. mRNA is dark line, and gRNA is light line. Middle panel: The two-cleavage, two-ligation model. Right panel: A model involving two transesterification reactions and two ligations. Adapted from Sollner-Webb (1991).

RNA INTERFERENCE

Some mRNAs never make it to the cytoplasm to be translated. They are degraded in the nucleus by a mechanism that seems to be also genetically regulated. This degradation occurs only when RNA is double stranded and is called RNA interference (RNAi). RNAi was suspected in experiments where transfection of the sense and antisense transcripts in cells had the same results. Usually the antisense sequence would temporarily block the sequence that corresponds to the sense transcript, resulting in silencing and gene regulation. If both sense and antisense transcripts have the same results inhibition is mediated by the formation of double-stranded RNA, which is the same in both cases. Subsequently, RNAi was found to block gene expression in *Caenorhabditis elegans* and other animals. In *C. elegans*, RNAi is genetically controlled, and it seems that it has been developed as a means of transposon silencing. In several RNAi mutants, there is mobilization of transposable elements in the germ line. For example, Tc1 transposon is active in somatic cells but not in the germ line. However, in mut-7 mutants (a protein with homology to Rnase D), Tc1 is also activated in the germ line. It seems, therefore, that, when double-stranded RNA is formed, an RNase specifically degrades it. RNAi may play an important role in protecting the genome from instability caused by the accumulation of transposons and repetitive sequence. Also, it may represent an ancient response to viral infections. Viral infections could result in the formation of double-stranded RNA. In fact, the antiviral effects of interferon are based on degradation of double-stranded RNA.

CHAPTER NINE

Compartmentalization of Transcription

—— – – –

PRIMER The spliced and modified mRNA must be transported to the cytoplasm for translation that will result in protein synthesis, which is the ultimate goal of transcription. The question addressed in this chapter involves the degree to which the topology of transcription and movement of the transcripts toward the nuclear pores is organized. This area of inquiry is a relatively new, and experiments are possible only by using modern image analysis techniques. Evidence is presented that transcription and movement of transcripts are organized topologically in the nucleus. Such an organization is compared with DNA replication. This information is very useful in understanding the spatial employment of these events.

So far, we have covered areas from the organization of DNA in the chromosomes to replication and the different steps of transcription. We have done all this with the three-dimensional aspects of molecular biology in mind; therefore, our study was oriented toward precise and organized structures. We have seen how precisely the splicing apparatus is folded to create unique structures and arrangement for the removal of the introns. Such structures are correlated to functions and also have an evolutionary history to account for them. What is, however, the picture when we move ourselves from the small scales to the larger ones? Are all the centers of transcription organized? Are all the splicing centers organized? Is their location in the nucleus determined or unique? In other words, are the phenomena that we have described (replication, transcription, splicing, and even transport of the mRNA) organized? To answer these questions, experiments should be designed where single transcripts can

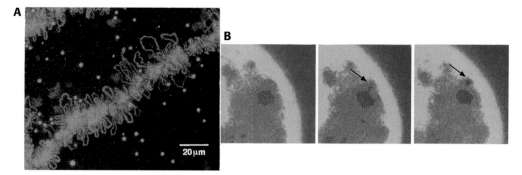

Figure 9.1. A: Splicing factors shown as green granules are placed close to loops of the lampbrush chromosome, which represents the active sites of transcription. **B:** Splicing factors (granule indicated by an arrow) move away from the storage (larger dot). J. Gall and D. Spector, Science 276: 1495–6 (1997). Reprinted with permission from American Association for the Advancement of Science.

be visualized. These experiments will require the use of very specific probes and sophisticated detection techniques. These techniques and the accompanying equipment are now available. The use of fluorescence in situ hybridization (FISH) and detection of the signal with advanced image analysis and confocal microscopy have provided interesting images of single transcripts and other processes in the nucleus.

SPLICING FACTORS

The first indication that transcription is compartmentalized was presented by the expression of splicing factors in the nucleus. It was found that splicing factors are not randomly distributed in the nucleus but that they are found in certain places or storage sites. Also, it has been determined that these factors are transported from these storage sites to transcriptionally active sites and occur in granules. Such distribution and movement of splicing factors can be seen in Figure 9.1.

VISUALIZATION OF SINGLE TRANSCRIPTS

As mentioned earlier, visualization of single transcripts requires several oligonucleotide probes that will hybridize to different areas of the transcript, conjugation of flourochromes on each oligonucleotide probe enabling the calibration of the fluorescent output, the digital imaging of a series of sections, and the processing of these images with the application of algorithms to restore out-of-focus light to its original points of origin. Such processes filter the diffuse pattern of hybridization seen in conventional techniques. Figure 9.2 is the product of those experimental manipulations that result in the visualization of single transcripts.

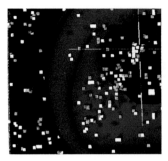

Figure 9.2. Observation of single (each pixel) beta actin transcripts. The hybridization probe was synthesized from the 3′-UT of the mRNA. The cross is the transcription site. Red is the boundary of the nuclear envelope, and we are looking from the bottom of the cell upward into the nucleus. Note that the flow of the transcripts in relation to their movement from the center outward is somewhat organized and forms a distinguishable path. R. Singer, Science 280: 585–90 (1998). Reprinted with permission from American Association for the Advancement of Science.

By using probes from different parts of the transcript labeled with different fluorescent dyes, we can observe a particular transcript as it is synthesized as well as its particular position in the nucleus or the cytoplasm in relation to other transcripts. For example, in Figure 9.3, we can observe the transcripts for beta actin (green) and gamma actin (red) in the nucleus (Figure 9.3A) and in the cytoplasm (Figure 9.3B). Note that individual beta and gamma actin mRNAs segregate independently in the cytoplasm.

We can also observe different regions of the transcript by using probes from, say, the 5′, 3′, or the splice junction (in our example using the beta actin transcript). When this is done, we can visualize the transcript in pieces, each one representing a particular area of it. Furthermore, such a procedure can result in a high-resolution dynamic view of a particular mRNA transcription site. Using such analysis, we can generate a snapshot of a particular gene in transcription. In the example with beta actin, such an experiment revealed 23 nascent RNAs in the process of elongation, with some of them (8) progressed through the distal 3′-UTR with 5 past the polyadenylation site. Experiments using more specific probes in relation to the cleavage or polyadenylation site might be able to distinguish whether the transcript is nascent, cleaved but undergoing polyA addition, or ready for transport (Figure 9.4).

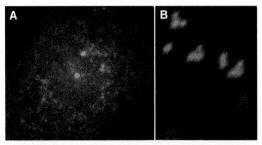

Figure 9.3. Nuclear (A) and cytoplasmic (B) localization of beta (green) and gamma (red) actin. The probes were both from the 3′-UT region. R. Singer, Science 280: 585–90 (1998). Reprinted with permission from American Association for the Advancement of Science.

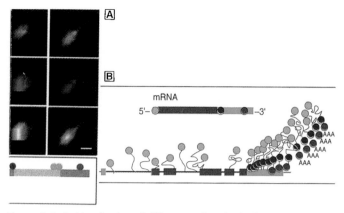

Figure 9.4. A: Visualization of different regions in the beta actin transcripts. The colors indicate the location of the probes. Red is from the 5'-UTR, blue is from the 3'-UTR, and green is from the splice site. In this experiment, the regions represented by blue and green were not resolved, so the co-localized voxels of blue and green are white. The distance from blue to red is nearly 490 nm, which is consistent with a linear RNA. **B:** A profile of transcription for the beta actin gene. This profile is generated with 23 nascent RNAs in the process of elongation. R. Singer, Science 280: 585–90 (1998). Reprinted with permission from American Association for the Advancement of Science.

THE MOVING OF TRANSCRIPTS OUT OF THE NUCLEUS

Other experiments have convincingly shown that transcripts move in an organized fashion along tracks toward the nuclear periphery. This movement might also indicate that transport could be coupled with transcription. In Figure 9.5, we can observe such tracks of the Epstein-Barr virus (EBV) transcripts from two different infected cell lines.

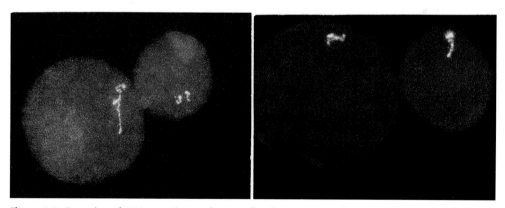

Figure 9.5. Detection of EBV transcripts tracks extending from an internal genomic site into the nuclear periphery. In the right part of the figure, note the accumulation of transcripts near the outer edge of the nucleus. The two figures show the tracks detected in two different cell lines transcribing EBV. S. Singer, Cell 57: 493–502 (1989). Reprinted with permission from Elsevier Science.

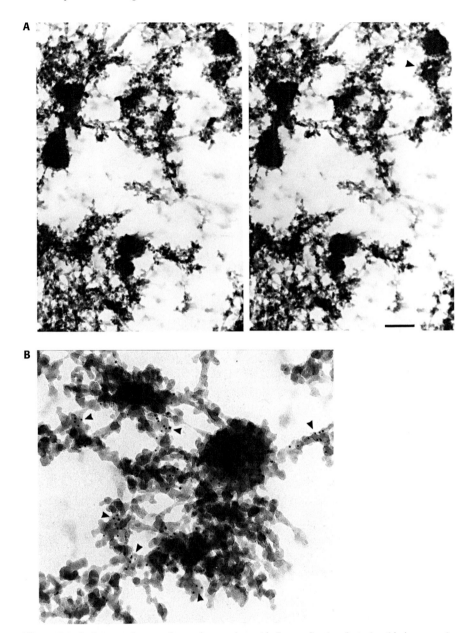

Figure 9.6. A: A stereo image of a nuclear region with five replication factories (dark masses). The nucleus has been incubated with biotin 11-UTP (for 2.5 minutes), and incorporation was visualized immunologically with gold particles. **A:** The labeling is very intense, and the particles are not visible. **B:** The particles in a factory and also in the adjacent chromatin (arrowheads) are visible. This display was captured after incubation for 60 minutes; consequently, it seems that replication occurs in the factories and then spreads to the adjacent chromatin. P. Cook, Cell 73: 361–73 (1993). Reprinted with permission from Elsevier Science.

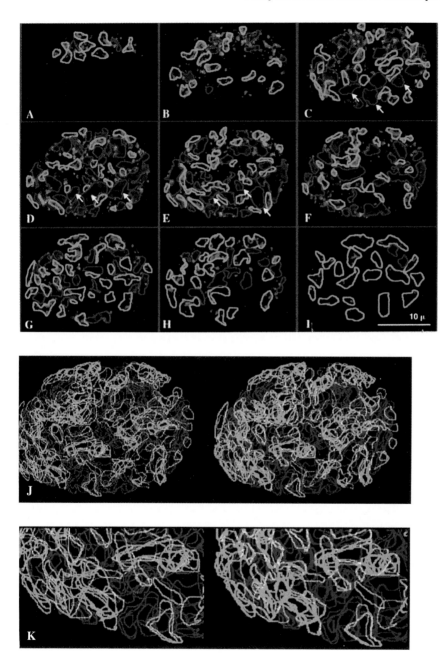

Figure 9.7. **A, B, C, D, E, F, G, H,** and **I:** Series of optical sections of mouse fibroblast nuclei (in early S phase) labeled for replication (green) and transcription (red). The arrows indicate the three nucleolar transcriptional sites. Note the apparent segregation of the clusters in groups. **J:** A stereo image of the reconstructed replication (green) and transcription (red) sites. **K:** An enlargement of the lower part of J. Note the zones of replication and transcription that are composed or group of factories. R. Berezney, Science 281: 1502–4 (1998). Reprinted with permission from American Association for the Advancement of Science.

SHUTTLING mRNA-BINDING PROTEINS

Some RNA-binding proteins called heterogeneous nuclear ribonucleoproteins (hnRNPs) play a role in the function of mRNA as it is shuttled to the cytoplasm from the nucleus. Other non-hnRNP factors also play a role as the mRNA is shuttled out of the nucleus. Both of these shuttling factors bind the mRNA as it is synthesized, usually on regulatory elements near the 3′ end. The factors are still bound to the mRNA in the nucleus as they are when the mRNA is circularized by the action of initiation factors and polyA-binding protein (see Chapter 10). At that point, another cytoplasmic factor binds and the whole complex seems necessary for mRNA localization, translation, and mRNA turnover. Then, the shuttling factors dissociate and go back in the nucleus, while the cytoplasmic factor is recycled.

SEGREGATION OF REPLICATION AND TRANSCRIPTION

As we have already seen, some events of transcription are topologically well organized in the nucleus. This organization becomes even more impressive when we think of how complex DNA packing is in the nucleus. Even though DNA is packed by higher order coiling, it is also tangled as a result of random bundling. Yet DNA must untangle to undergo replication and transcription. Is there an order in this tangle, especially when it comes to replication and transcription? We do know that replisomes do not act alone, but rather act in unison. A large group (some times hundreds of them) make up a replication factory. These factories have been observed fixed on a nucleoskeleton (Figure 9.6).

Transcription machines are also concentrated in analogous sites, or factories, but in contrast with replication, transcription takes places continuously. Also, it seems that replication factories replace transcriptional factories. To make sense of all this information, it is important to see where these factories are located. For example, can a transcriptional factory be a replication factory at the same time? Are the factories, in fact, segregated in large areas in the nucleus, or they are intermixed? Modern microscopy techniques have shown an amazing high-order arrangement of genomic function in the nucleus. Replication and transcription are segregated in higher order domains, which means that many factories of transcription or replication segregate to create distinct zones in the nucleus. Obviously, such results indicate that a group of replication factories must be decommissioned together after DNA duplication to become transcription factories. We can see this higher order of arrangement and cluster distribution of replication and transcription sites in nuclei subjected to optical sections (Figure 9.7).

CHAPTER TEN

Protein Synthesis

——— – – –

—————————————

PRIMER When the mRNA is in the cytoplasm, it is recognized by factors and complexed with the ribosome, which is the place where the message is decoded and the protein is built up. We are rather fortunate because most of the players participating in protein synthesis are known at the 3-D level; thus, we can reconstruct all the steps of protein synthesis at the 3-D level. Some of the figures, especially the ones depicting the process of protein synthesis with the ribosome, are so stunning that I virtually feel I am there viewing the process first hand. For this spectacle to flow in the most comprehensive way possible, I selected a certain sequence for the presentation. First, I present the 3-D structure of the two major players in protein synthesis: tRNA (alone and complexed with aatRNA synthetase) and ribosome. The reader must become very familiar with both, especially with the ribosome. Even though we know the structure of the ribosome in high resolution, I first introduce its structure by showing a series of images from low to high resolution. This approach should enable the reader to understand the structure better. Having introduced the basic players, I proceed with the first stage of protein synthesis, the initiation. This is compared in both prokaryotes and eukaryotes, and the known 3-D structure of initiation factors and their interaction with rRNA are presented. The second step in protein synthesis, the elongation, involves the placement of tRNA (by interaction with elongation factor) in the A-site of the ribosome and its movement to the P-site, after peptide bond is achieved, and finally its exit from the ribosome. This movement of tRNA from the A- to P- to E-site is presented with beautiful 3-D images.

We then examine interactions between tRNA and the components of the ribosome, that is the 16S and 23S rRNA. For this, the structures of 16S and 23S rRNA are examined, and the important areas of interaction are highlighted. Having become accustomed to these structures and inter-actions, we can visualize the codon-anticodon interactions as well as the decoding center of the ribosome that catalyzes the peptide bond. This jour-ney through the events of protein synthesis triumphantly shows how the structure reveals the function. Finally, the termination process is presented. Here again the structure of the termination factors provides important insights of the mechanism of termination.

I often say that viewing the process of protein synthesis at the 3-D level is so beautiful that I would not mind living in a ribosome!

The final goal of all the processes discussed so far is to decode genetic information. We know this to be true because the nucleotide sequence in the genes contains the information that can be used to synthesize proteins. This decoding is called translation. Perhaps the process of protein synthesis is the most complex yet. For nature to achieve this process, it should devise a plan that would make use of the DNA sequences as codes to create proteins. When the structure of DNA was revealed in 1953, scientists were faced with this monu-mental problem. They knew that because the DNA is the genetic material that contains the information for protein synthesis, there must be an association between DNA sequence and amino acids, which are the building blocks of pro-teins. It did not take long to figure out that a combination of three nucleotides is the code for an amino acid. Because we have only four bases in the DNA, the only combination that would match closely the number of amino acids is 3 nucleotides. Based on this discovery, we can have 64 different combinations of 3 nucleotides that code for 20 amino acids. These 64 combinations, or codons, constitute the genetic code. Because there are 20 amino acids, some amino acids are coded by more than one codon. In such cases, the first two nucleotides of the different codons are usually the same, and the third (wobble nucleotide) is different. In all, the genetic code has 61 codons accounting for amino acids and three codons that signal the end of protein synthesis and are called stop codons.

Having deciphered the genetic code, researchers intensified their search for the mechanisms of translation. Because protein synthesis occurs in the cyto-plasm, it became obvious that the DNA must be copied in and the uninter-rupted coding sequences must travel to the cytoplasm. The discovery of mes-senger RNA heralded the beginning of what is known now as gene regulation (transcription, splicing), which is also the main focus of this book. Once in the cytoplasm, the mRNA is recognized and placed in the ribosome where decoding and protein synthesis occurs. Every triplet must be recognized and decoded, leading to peptide synthesis. How does this recognition of a codon become possible? Nature invented an amazing trick: a new type of nucleic acid that can recognize the triplet (by complementary sequences) and, at the same time, be charged with an amino acid that corresponds to the particular codon. This new player is the transfer RNA (tRNA). The complementary sequence of tRNA is called anticodon. However easy this might look now, the association of a particular amino acid with a tRNA that recognizes the corresponding codon for this amino acid must have taken a few evolutionary tricks that we have not yet been able to picture clearly. Nevertheless, the evolution of tRNAs as decoding machines was one of the most spectacular discoveries leading to the establishment of a true genetic material and the onset of cells and organisms.

Therefore, the most important players in protein synthesis are the tRNA and the ribosome. They both collaborate for the decoding process and the synthesis of proteins. We are fortunate enough that most of the factors and players involved in protein synthesis have been well studied and that their 3-D structures have been solved. These findings have provided us with spectacular views of the event of protein synthesis. Before we proceed to the different stages of protein synthesis, we should become familiar with the 3-D structure of the major players, which are tRNA and the ribosome.

THE MAJOR PLAYERS IN PROTEIN SYNTHESIS

The 3-D Structure of tRNA

The tRNA is made up of about 75 nucleotides that fold into a secondary cloverleaf structure. This secondary folding is possible as a result of the formation of stems (or arms) due to complementary sequences and unpaired sequences that form the loops. The folding brings together the 5′ and 3′ end. The 3′ end sequence CCA is common in all tRNAs and constitutes the acceptor end that is charged with the amino acid. There are four stems and three loops in all tRNAs – the acceptor stem, the D-stem and D-loop, the anticodon stem and loop, and the TψC stem and TψC loop. The D-stem and -loop are so called because of the presence of modified uracil bases (dihydrouracil). The TψC loop contains a modified nucleoside, pseudouridine (ψ), flanked by the invariant T and C. Only 15 invariant bases in tRNAs are necessary for folding into the three dimensions (Figure 10.1). Also, in some tRNAs, another loop (variable loop) is present between the anticodon and TψC stem. In fact, tRNAs contain many additional modified nucleosides, some of which are methylated. The modifications in the tRNAs are made after transcription by multiple enzyme systems. The modifications seem to be necessary for tRNA function. Such a conclusion has been derived from experiments where synthesis of tRNAs with unmodified bases resulted in its inability to bind amino acids.

As mentioned earlier, the invariant bases seem to play a significant role for the three-dimensional folding of the tRNA. In particular, there are extensive interactions between the invariant bases of the D-loop and the TψC loop.

A

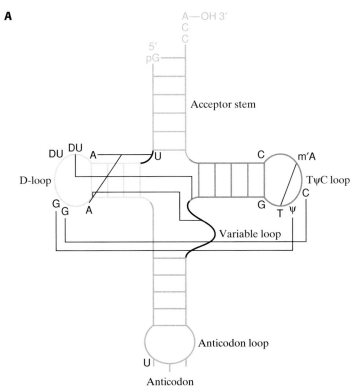

Figure 10.1. A: The secondary structure of tRNA showing the stems and the loops. The invariant bases and the two dihydrouracils are presented. The lines connect bases that interact to fold the t-RNA into its characteristic three-dimensional structure (B). Note the stacking of the anticodon bases (arrows). E. Westhof et al., PDB file (IMB Jena) TRNA09, Acta Crystallogr. 44: 112–23 (1988).

These interactions fold the tRNA in its characteristic inverted L-shaped three-dimensional structure (Figure 10.1). The folding involves numerous extraordinary pairings (not the normal Watson-Crick base pairing). Another interesting feature is the structure of the anticodon. The three bases are stacked and project out to the right away from the backbone of the tRNA. In fact, the anticodon backbone is twisted into a partial helix shape. These structural features of the anticodon facilitate its interactions with the codons in the mRNA.

The Charging of tRNA with Amino Acid

The major player for charging tRNA with amino acid is the enzyme aminoacyl-tRNA synthetase. Each enzyme (there are as many such enzymes as amino acids) recognizes only its cognate tRNA and attaches its corresponding amino acid in the acceptor end. The enzyme has, therefore, three binding sites – the ATP, the amino acid, and the tRNA binding site. The reaction utilizes ATP, and the steps follow. First, the amino acid and ATP form aminoacyl-AMP. Then

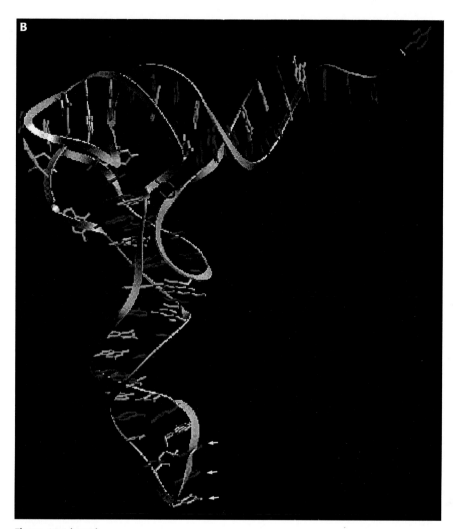

Figure 10.1(cont.)

tRNA binds the enzyme. Subsequently, tRNA is charged with the amino acid (Figure 10.2).

The 3-D Structure of *E. coli* Glutaminyl-tRNA Synthetase and Its Interactions with tRNA^{Gln}

The enzyme consists of four domains that make contacts with the tRNA. A portion of the protein is very similar to the so-called dinucleotide fold. This structural motif is found in dehydrogenases, kinases, synthetases, and proteins that utilize ATP or GTP. The first half of this domain is made up of beta strand 1, 2, and 3 and alpha helix G and provides the major contacts with the ATP substrate. The second half, consisting of beta strands 9 and 10 and alpha helix

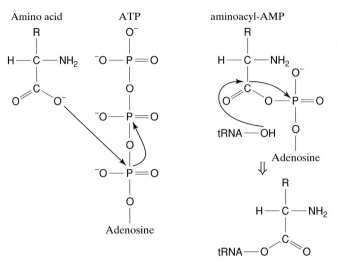

Figure 10.2. The steps resulting in the charging of a tRNA with an amino acid by aminoacyl-tRNA synthetase.

H, makes contacts with the glutamine substrate and with the tRNA acceptor stem. The acceptor end binding is from the amino terminus of helix D to the carboxyl end of beta strand 8 (Figure 10.3).

Let us now see the interaction of the enzyme with the acceptor stem and end to obtain some insights about the specificity of the enzyme. As pointed in the structures presented in Figure 10.3, the enzyme disrupts U1:A72 base pairing. We can see these interactions in Figure 10.4. Leu-136 is the amino acid that is inserted between U1 and A72 resulting in disruption of the base pairing. A neighboring Arg-133 forms hydrogen bonds with both A72 and G73.

Obviously, the specificity of tRNA synthetases must be associated with the preceding structural interactions. It has been suggested that base 73 is the discriminator base. tRNA with A in that position tends to code for hydrophobic amino acids, while tRNAs with a G in position 73 code for polar amino acids. Therefore, the identity of this position seems to have played an important role in the evolution of and specificity for tRNA recognition. These ideas have been supported by results where tRNAs mutated at position 73 tend to change their specificity for amino acids. Further support of the implication of sequence and structure of the acceptor end in the specificity of the synthetases comes from the 3-D structures of different tRNAs. If the 3-D structures of two different tRNAs are superimposed, we can see that they align very well except for the regions that contact the synthetases, namely the anticodon loop, the D-loop, and the acceptor stem and end (Figure 10.5). Such differences might account for the specificity of aminoacyl-tRNA synthetases for their cognate tRNAs. It is believed that tRNA binding to aa-RSs is processed in two steps – an initial one with broad specificity followed by a more precise recognition that involves

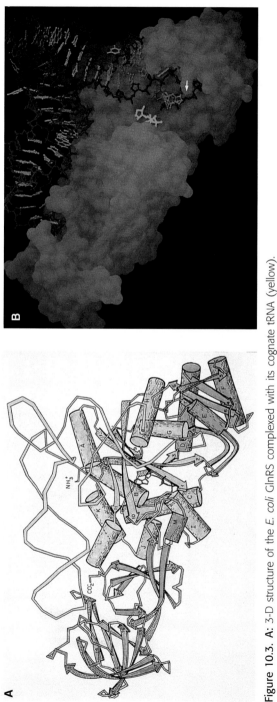

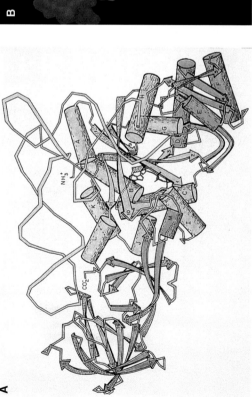

Figure 10.3. A: 3-D structure of the *E. coli* GlnRS complexed with its cognate tRNA (yellow). The different domains are denoted with different colors. ATP (blue) is shown bound by the dinucleotide motif. Other contacts of the enzyme and the tRNA are with the anticodon and the D-loop. **B:** Same structure as in A, but the GlnRS is presented as a surface representation to better indicate the interactions between the enzyme and the tRNA. Note the deep cleft where the ATP (green) and the acceptor end are buried. The anticodon and the D-loop also interact with protein motifs that create clefts. Also note the part of the protein that is inserted between the 5' and 3' ends of the tRNA and disrupts the U1:A72 base pair (arrow). T. A. Steitz, Science 246: 1135–42 (1989). Reprinted with permission from American Association for the Advancement of Science.

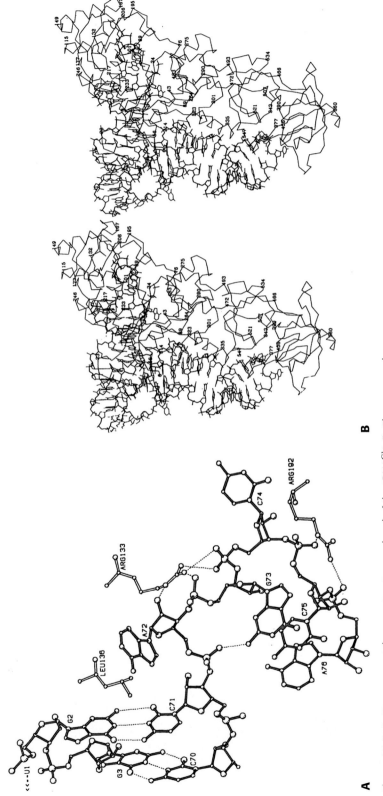

A

B

Figure 10.4. A: Sequences at the acceptor stem and end of the tRNA^{Gln}. **B:** The same end as in A, but it is presented in the 3-D structure. Note the insertion of Leu-136 into the β structure. Note the insertion of Leu-136 (part of the beta hairpin helix E and beta strand 5) between U1 and A72 and the interaction of the acceptor end with Arg-133 and Arg-192. The G2 : C71 and G3 : C70 pairing is recognized by the beta loop (connecting beta strands 6 and 7) and helix H (see also Figure 10.3). T. A. Steitz, Science 246: 1135–42 (1989). Reprinted with permission from American Association for the Advancement of Science.

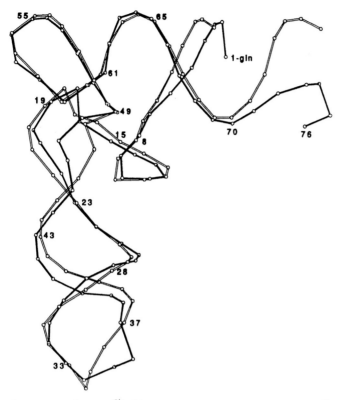

Figure 10.5. The tRNA^{Gln} (black) superimposed with the yeast tRNA^{Phe}. Note the major differences in the acceptor end and the anticodon loop. T. A. Steitz, Science 246: 1135–42 (1989). Reprinted with permission from American Association for the Advancement of Science.

changes in the conformation of both tRNA and the enzyme. It is true that bound noncognate tRNAs are usually not aminoacylated.

The aminoacyl-tRNA synthetases are grouped as class I and class II. There are three major differences between class I and class II aa-tRNA synthetases. First, even though the overall structure is similar, there are differences. In class I (see the structure for GlnRS as described earlier), the catalytic site (ATP binding) is made up of five parallel beta strands. In class II, the catalytic site is made up of six antiparallel beta strands. Also, in class I the anticodon binding domain is made up exclusively of beta strands, but this domain is a combination of beta strands and alpha helices in class II enzymes. Second, class I enzymes approach the acceptor stem from the minor groove, while class II enzymes approach from the major groove. Third, class I aminoacylates the 3′-OH of the terminal adenosine, while class II aminoacylates the 2′-OH. However, exceptions to this rule have been found. The cysteinyl-tRNA synthetase from the archeon *Methanococcus jannaschii* and *Deinococcus radiodurans* lacks the characteristic sequence motifs found in all the other members of the two classes.

Enzymes from yeast and higher eukaryotes differ from their corresponding prokaryotic ones in that they have an extension in the N- or C-termini.
Aminoacyl-tRNA synthetases can be monomers, dimers, and sometimes
tetramers. However, in eukaryotes, nine enzymes can be associated together,
and this complex also contains additional polypeptides. This association appears
to be possible through the additional domains found in the N- and C-termini.
In eukaryotes, maturation and aminoacylation of tRNAs occurs in the nucleus.
After these processes have taken place, the charged tRNAs are exported
through the nuclear pore by a complex of a carrier, exportin t, with Ran-GTP.
In the cytoplasm after hydrolysis to Ran-GDP, the exportin t is dissociated, and
the charged tRNA is ready for translation.

The Ribosome
The mRNA and the charged tRNAs meet at the ribosome. The ribosome
is the place or rather the factory where proteins are produced by decoding
the mRNA. The ribosome is a very complex structure that eventually accounts
for interactions between its components, mRNA, and tRNAs. All these interactions that lead to protein synthesis will be described in details later, but
now some general features of the ribosome will be presented to familiarize the
reader with the basic structure of the ribosome. I urge the reader to become
familiar with the 3-D structure of the ribosome, inhabit it, and become a specta-
tor of the marvelous play of proteins synthesis. The ribosome has two subunits,
the large and the small, and is made up of RNA, which is called ribosomal
RNAs, and proteins. The size of the rRNAs and the number of the proteins
depend on the species origin of the ribosome. Bacterial ribosome has a sedi-
mentation factor of 70 and is designated as 70S. The large subunit (50S) contains
the 23S rRNA (nearly 2,900 bases) and the 5S (120 bases). It also contains nearly
30 proteins. The small subunit (30S) contains the 16S rRNA (nearly 1,500 bases)
and 21 proteins. The eukaryotic ribosome (80S) is somewhat larger. The mammalian large subunit (60S) contains the 28S rRNA (nearly 4,700 bases), the
5.8S rRNA (160 bases), the 5S rRNA (120 bases), and 49 proteins. The small
subunit (40S) contains the 18S rRNA (nearly 1,900 bases) and 33 proteins.
The different proteins are discriminated by the letter L for large subunit or
S for small subunit. The ribosomes are built as subunits (large or small) in
the nucleolus and then assembled in the cytoplasm. Maturation and transport
of the ribosomes require three proteins – Noc1p, Noc2p, and Noc3p. These
proteins can be isolated in complexes of two. The Noc1p/Noc2p complex associates with the 90S and 66S preribosomes and is found in the nucleolus, while
the Noc2p/Noc3p complex associates with the 66S preribosomes and is mainly
found in the nucleoplasm. An important characteristic of ribosomal proteins is
that they contain a globular domain and an extended element that is believed
to play a role as an anchor and stabilize the structure of the ribosome (see later
sections).

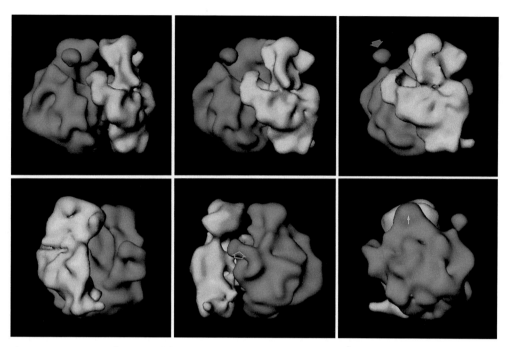

Figure 10.6. A series of images of the prokaryotic (*E. coli*) ribosome at 25 Å. The large subunit is blue, and the small subunit is yellow. Starting from top left, where we see the ribosome from the E-site, we go around it to examine its features from different angles. In the lower left, for example, we see the ribosome from the other site, which is the A-site. The L1 protein (thick arrow) is at the exit site, and the stalk (open arrow) that contains the L7/L12 protein complex and interacts with factors and tRNA is at the A-site. Note that the interface between the two subunits is basically a passage that creates a domelike structure. In this structure, the tRNAs bind and move; the peptide bond is also made here. The central protuberance (CP) is where the 5S rRNA is located and is shown by a thin arrow. J. Frank, Amer. Scientist 86: 428–39 (1998). Courtesy of Dr. J. Frank.

Most of the studies on the 3-D structure have been performed with the prokaryotic ribosome. Initial studies used cryoelectron microscopy and image reconstruction to decipher the structure of the ribosome. Recently, however, X-ray crystallography images have emerged. In Figure 10.6, some 3-D images of the ribosome are rotated so that the reader can obtain the whole view of the ribosome and become accustomed with its major features. The two subunits form an interface, which can accommodate the tRNAs as they move from one side of the ribosome to the other. The aa-tRNA (say $n+1$) enters the A-site (A for aminoacyl). Next to it is the number n tRNA, which occupies the P-site (P for peptidyl) and carries the polypeptide. The E-site (E for exit) is at the other end of the interface. Here the deacylated tRNA leaves after it has transferred the peptide from the P-site to the A-site tRNA. In the next pages, the orientation of the ribosome might differ from that presented in Figure 10.6, depending on what is shown. Sometimes this change of view is necessary for better clarity.

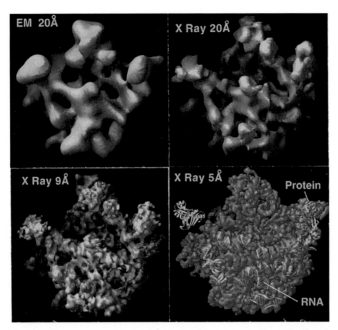

Figure 10.7. 3-D structures of the large subunit as seen from the interface (imagine that you are the small subunit and you look at the large one). As the resolution becomes higher, we can see the fine details of the structure. Eventually, when we reach a resolution of 5 Å or less, we can see proteins and rRNA fitted in the structure. Note that the overall structure of the ribosome has not changed, but the higher resolution has helped us to understand the interactions of the players of protein synthesis and decipher the mechanisms involved. J. Frank and T. A. Steitz, Science 285: 2050–1 (1999). Reprinted with permission from American Association for the Advancement of Science.

In the next pages, the structure of the ribosome and the events of translocation will be presented in more detail, and the reader will be able to visualize the process of protein synthesis. The ribosome structures in Figure 10.6 are revealed at the resolution of 25 Å, which is possible by cryoelectron microscopy. Let us see, however, for clarity reasons, how the overall structure of the ribosome has improved with better resolutions (Figure 10.7).

Having described the major players in protein synthesis, we will now proceed to examine the process in detail. The process of protein synthesis has three steps. During initiation, the mRNA and the initiation codon are recognized and interact with the small subunit. During elongation, the actual protein synthesis occurs. At termination, stop codons are recognized, resulting in the termination of the polypeptide synthesis.

INITIATION

The major goal of the initiation process is the binding of mRNA in the small subunit and the recognition of the initiation by the initiator tRNA. It is imperative that the mRNA binds on the correct site of the ribosome with its initiation codon at the correct place. Despite the fact that this goal is the same in prokaryotes and eukaryotes, the initiation process involves several steps and factors that are not the same in prokaryotes and eukaryotes. Therefore, we will examine the process separately.

Initiation in Prokaryotes

The initiation codon in prokaryotes is usually AUG with a frequency of more than 90%. This is the codon for methionine. However, UUG and GUG can also be used as initiation codons, albeit rarely, because the specific tRNA^{Met} that binds the initiation codon, formylated Met-tRNA^{Met}, can also recognize these two other codons. The regular tRNA^{Met} that is used to decode internal methionine codons cannot recognize them. The formyl group is incorporated in the amino group of methionine.

The first step in the initiation is the dissociation of the 70S ribosome, which is achieved by the initiation factor 1 (IF-1). To prevent reassociation, another initiation factor, IF-3, binds to the 30S subunit. Subsequently, IF-1 and IF-2 bind alongside IF-3. IF-2 is the factor that promotes binding of the fMet-tRNA^{Met} to the 30S initiation complex. IF-3 promotes binding of mRNA to the 30S subunit and cooperates with IF-2 to ensure that only the initiator tRNA binds the P-site. The mRNA is positioned at the correct place due to complementarity of sequence AGGAGGU just upstream the initiation codon with the 3′ end of the 16S rRNA UCCUCCA. The sequence AGGAGGU is called Shine-Dalgarno or SD after its discoverers. After the 30S initiation complex has been formed, the large subunit binds with loss of IF-1 and IF-3. The last step involves dissociation of IF-2 with hydrolysis of GTP to GDP. The 70S initiation complex is now ready to begin elongation (Figure 10.8).

Initiation in Eukaryotes

There are many differences and similarities between prokaryotic and eykaryotic initiation of protein synthesis. In eukaryotes, there is no SD sequences for the positioning of mRNA on the ribosome's small subunit. Instead, the cap is recognized by specialized initiation factors. The methionine that is bound to the tRNA responsible for recognizing the initiation codon is not formylated. Instead, the initiator tRNA is modified by a 2′-OH phosphorylation at base 64. The initiation factors, designated as eIF (for eukaryotic), occur more frequently in eukaryotes, but some of them are similar and perform similar functions with their prokaryotic counterparts.

After dissociation of the ribosomal subunits, eIF-3 binds the 40S subunit and prevents reassociation. Likewise, eIF-6 binds the 60S subunit to prevent reassociation with the small 40S subunit. Next, eIF-2 binds the initiator tRNA and brings it to the small subunit. For eIF-2 to bind the initiator tRNA, eIF-2 : GDP must become eIF-2 : GTP. This is catalyzed by eIF-2B. Then, this complex must

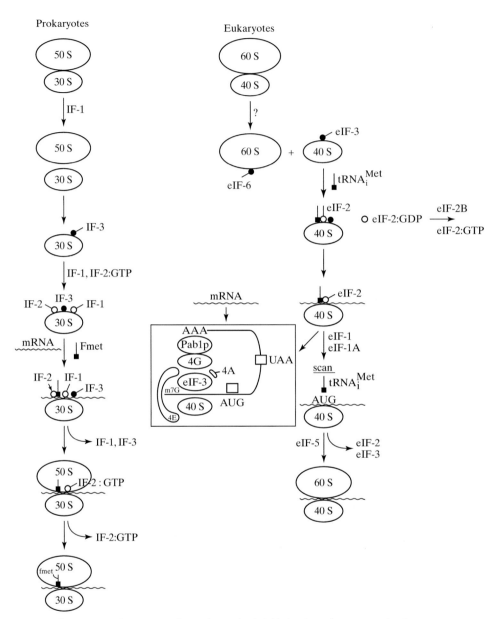

Figure 10.8. The process of protein synthesis initiation in prokaryotes and eukaryotes.

be recruited to mRNA. The major factor involved in this step is eIF-4F, which is composed of three parts: eIF-4E, eIF-4A, and eIF-4G. eIF-4E is the actual cap-binding domain. EIF-4G binds eIF-4E, eIF-3 (bound to 40S subunit), and the polyA-binding protein, Pab1p. These interactions will help the 40S sub-unit not just to bind the mRNA but also to circularize it. The eIF-4A, the other part of eIF-4F, is a member of the DEAD family (see Chapter 7) with heli-case activity. This factor unwinds hairpins usually found in the 5′ leaders of eukaryotic mRNAs. This unwinding activity of eIF-4A is supported by eIF-4B, which stimulates the binding of eIF-4A to the mRNA. eIF-4B requires ATP to complete its function.

As soon as the cap is recognized, the ribosome scans the mRNA until it finds an AUG, the initiation codon. The best context for this recognition is the sequence ACCAUGGG. eIF-1 and eIF-1A are involved in this process. In nearly 10% of the cases, the ribosome will bypass the first AUG. After this, eIF-2 and eIF-3 are released, and the small and large subunits are reassociated with the help of eIF-5, which is a GTPase (Figure 10.8).

However, cases of cap-independent translation have been noted. In particular, sequences that provide high-affinity binding sites for binding to eIF-4G, eIF-3, or the 40S subunit have been found in mRNAs. These sequences, which are called internal ribosome entry sites (IRESs), were first discovered in viral mRNAs and then in cellular mRNAs. These might have been evolved as a strategy for viruses to avoid the general inhibition of the translation initiation step occurring during the G2/M phase of the cell cycle. This inhibition is the result of the loss of the eIF-4E ability to bind the cap structure.

The Structure of IF-1

The structure of IF-1 (from *Thermus thermophilus*) has been solved as complexed with the 30S ribosomal subunit. The structure shows the characteristic OB fold found in many ribosomal proteins that belong to the S1 family. The fold consists of a barrel of five beta strands. IF-1 binds to the 30S ribosomal subunits in a cleft formed between H44, the 530 loop, and protein S12 (for the position of these elements see later sections). The residues that interact with the ribosome are basic; however, acidic residues face the solvent surface. This asymmetric distribution of charge could be important in stabilizing the binding of IF-1 to the 30S subunit. When IF-1 is bound to the small subunit, there are considerable alterations in the whole structure of the 30S subunit. A loop from IF-1 inserts into the minor groove of H44 and flips bases A1492 and A1493 (Figure 10.9A). Also, binding disrupts the canonical base pair between A1413 and G1487 (Figure 10.9B). Except for the change in H44 structure, the movements of the domains of 30S subunit are also small, the most notable of which is tilting of the head toward the body of the subunit. The location of IF-1 on the small subunit suggests that IF-1 would block tRNA binding to the A-site. This blockage might indicate that such binding ensures that a free aatRNA cannot bind the A-site during initiation. The importance of this region in the decoding process will become clear in the next sections.

The Structure of IF-3

The structure of IF-3 is also very revealing and instructive about its function. IF-3 is composed of two domains with exposed beta strands and is connected by an alpha helix (Figure 10.10). This structure suggests that IF-3 is able to bind and bridge two distant regions of the ribosome. Indeed, IF-3

A

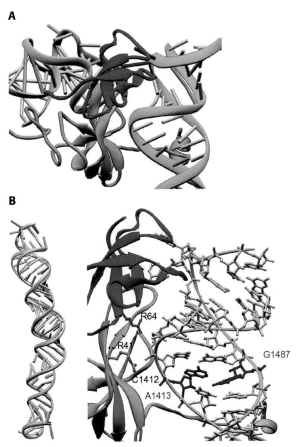

B

Figure 10.9. A: The structure and interaction of IF-1 with elements of the 30S ribosomal subunit. IF-1 is purple, helix 44 is cyan, the 530 loop is green, and protein S12 is orange. Note the interaction with H44 and the flipping out of A1492 and A1493 (red). Also note the characteristic OB fold with the antiparallel beta strands and the two helices. The S12 structure is also characteristic for ribosomal protein containing a globular domain and an extended segment. **B:** Changes in helix 44 with (blue) or without (yellow) IF-1 binding. Note the disruption in the A1413-G1487 base pairing. V. Ramakrishnan, Science 291: 498–501 (2001). Reprinted with permission from American Association for the Advancement of Science.

can be cross-linked to S7 and S11, two proteins that are far away from each other, on the opposite sides of the cleft. Also, the structure of IF-3 resembles what has been seen in other ribosomal proteins. Two RNA-binding domains connected by a helix is also characteristic of L9 protein. It seems that IF-3's location spans the region from the 50S subunit side of the platform to the neck of the 30S subunit. IF-3 is known to recognize the three G : C pairs of the anticodon stem, which are unique to the initiator tRNA. The C-terminal domain of IF-3 is the one most likely to be involved in recognizing the anticodon stem. However, IF-3 does not bind the anticodon itself, allowing it to bind the start codon. Such placement of IF-3, therefore, can explain its function in the

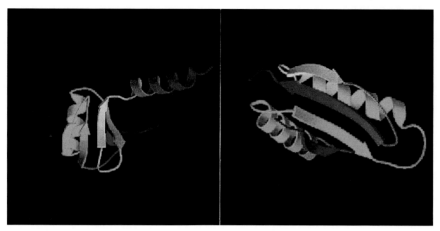

Figure 10.10. The N-terminal (left) and C-terminal (right) domain of the *Bacillus stearother-mophilus* IF-3. Note the beta-stranded exposed regions and the helix connecting the two domains. V. Biou et al., PDB file 1TIF & 1TIG, EMBO J. 14: 4056–64 (1995).

selection of initiator tRNA. The location of eukaryotic IF-3 on the 40S subunit and its structure are very different. eIF-3 is a big complex consisting of eight subunits with no homology to bacterial IF-3.

The Structure of IF-2/eIF-5B

Among all the different factors involved in initiation, only IF-2/eIF-5B and IF-1/eIF-1A are common in all three kingdoms. As mentioned earlier, IF-2 and eIF-5 are necessary to join the ribosomal subunits, and they function as GTPases. The 3-D structure of IF-2/eIF-5B has been solved and might explain how this factor promotes the joining of the ribosomal subunits. Remarkably, the overall structure looks like a chalice. The cup is made up of the G-domain (GTP-binding) domain II and domain III. The cup is connected to the base (domain IV) by a long alpha helix (Figure 10.11). The G-domain is very similar to the equivalent domains from EF-G and EF-Tu (see later). The structure suggests that the C-terminus (domain IV) interacts with the small ribosomal subunit. This interaction is also inferred by the fact that domain IV interacts with eIF-1A, which is known to bind the A-site. The N-terminal core (very conserved residues of the G-domain) binds to the large subunit; therefore, the whole protein plays the role of a bridge for the joining of the ribosomal subunits. The bridging most likely depends on the spatial arrangement of the G-domain and domain II in relation to domains III and IV. This is because, upon GTP and Mg^{2+} binding, domain II is rotated by 8 degrees toward the G-domain and domain III is rotated outwardly by 7 degrees.

The Structure of eIF-4E and eIF-4G

The major feature of the mouse eIF-4E structure is a large curved (concave) surface made up of antiparallel beta strands. The cap binds in a narrow

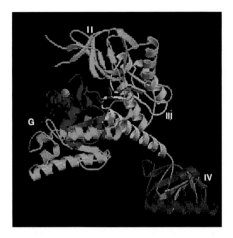

Figure 10.11. The 3-D structure of *Methanobacterium thermoautotrophicum* IF-2/eIF-5B. The domains are designated. Also shown is the binding site for GTP (red) and magnesium (gray sphere). See text for details. Roll-Mecak et al., PDB file 1G7T, Cell 103: 781–92 (2000).

binding slot on the concave surface. The recognition of the cap is mediated by base sandwiching between two conserved tryptophans. In addition to these two tryprophans, there are other interatomic contacts: hydrogen bonds or van der Waals contacts (Trp-102, Glu-103, and Trp-166), direct interactions with the ribose and the diphosphate moieties of the cap (Trp-56, Arg-157, and Lys-162), and water-mediated contacts (Lys-162 and Arg-112) (Figure 10.12). The convex

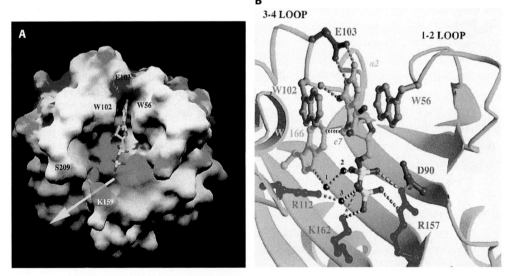

Figure 10.12. A: Cap (7-methyl-GDP) binding surface of eIF-4E, colored for electrostatic potential. Selected residues involved in recognition are labeled. The putative RNA path is indicated by a yellow arrow passing between Ser-209 and Lys-159. **B:** Details on the cap recognition, showing the interactions and contacts of protein residues with the 7-methyl-GDP. The interaction of Arg-157 is stabilized with a salt bridge (side chain ionic interactions). S. Barley, Cell 89: 951–61 (1997). Reprinted with permission from Elsevier Science.

ventral side of the protein contains many conserved hydrophobic residues. This region is suspected to be accessible for interactions with other initiation factors. The structure of a conserved portion of the human eIF-4GII is crescent-like and consists of ten alpha helices arranged as five HEAT repeats. The HEAT repeat is made up of two antiparallel alpha helices that occur in tandem arrays repeated 3 to 22 times. The HEAT repeat-containing proteins do not share any absolutely conserved amino acids. The structure also reveals that eIF-4G binds eIF-4A for cap-dependent initiation but can also bind IRES for cap-independent initiation.

The Structure of eIF-1 and eIF-1A

Both eIF-1 and eIF-1A are involved in the selection of the initiation site. After the 43S preinitiation complex is formed, these factors direct it from the 5′ end of the mRNA to the initiation codon. The structure of eIF-1 is similar to the OB domain found in many ribosomal and RNA-binding proteins (see structure of IF-1) and consists of two helices and a five-stranded beta sheet (Figure 10.13A). The N-terminus of eIF-1 is made up of an extended chain. eIF-1 does not interact with eIF-5, but it does bind to eIF-3, and this interaction might explain how eIF-1 is recruited to the 40S ribosomal subunit. eIF-1A has a similar structure and, with the exception of the OB domain, also contains an additional domain (Figure 10.13B). eIF-1A binds the single-stranded RNA sequence in a site-specific, but not sequence-specific, manner. This selectivity might indicate that this factor would rather interact with mRNA rather than

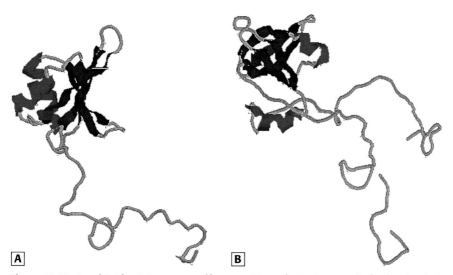

A   B

Figure 10.13. A and **B:** The 3-D structure of human eIF-1 and eIF-1A, respectively, showing their characteristic OB fold consisting of two helices (red) and a five-stranded beta sheet (blue). **A:** C. M. Fletcher et al., PDB file (IMB Jena) 2IF1, EMBO J. 18: 2631–7 (1999). **B:** J. L. Battiste et al., PDB file (IMB Jena) 1D7Q, Mol. Cell 5: 109–19 (2000).

tRNA or rRNA. The RNA-binding surface is a large area that encompasses the OB fold as well as a groove leading to the other domain. Indeed, mutations in the RNA-binding surface cause an inability of eIF-1A to assemble the preinitiation complex at the AUG codon.

ELONGATION

After the initiation codon is recognized, the process of protein synthesis enters the second phase, that of elongation. Let us now describe the basic events that take place during the elongation process. The initiator tRNA occupies the P-site in the ribosome and interacts with the AUG codon. The next mRNA triplet is in the A-site and is accessible to the next incoming tRNA. As soon as the tRNA number 2 is in the A-site, the number 1 amino acid (in this case Met), which is hooked in the tRNA, is transferred to the A-site, and a peptide bond is created with the number 2 amino acid. Then the tRNA from the A-site translocates to the P-site, and the process is repeated many times until a stop codon signals the termination of the synthesis of the peptide. The deacylated tRNA leaves the P-site through the E-site of the ribosome. The entry of aminoacyl-tRNA (aatRNA) to the A-site and the subsequent translocation are controlled by the so-called elongation factors. The entry of the aatRNA to the A-site is controlled by the EF-Tu (in prokaryotes) or eEF-1a (in eukaryotes). For the translocation from the A- to P-site, the factor that is needed is EF-G (in prokaryotes) or eEF-2 (in eukaryotes). Finally, EF-Ts (prokaryotic) and eEF-1bg (eukaryotic) are needed to replace GDP by GTP in EF-Tu and eEF-1a, respectively.

EF-Tu binds GTP and then is able to form a ternary complex with the aatRNA. The ternary complex can then occupy the A-site. After this has been accomplished, EF-Tu-GDP leaves the ribosome. EF-Ts replaces the GDP with GTP and the EF-Tu-GTP is again ready to bind another aatRNA. The peptide bonding in the A-site is catalyzed by the peptidyl-transferase center of the ribosome (see later for more details). The peptide bond is achieved by transesterification as shown in Figure 10.14.

The classical series of events that relate to the aatRNA ternary complex entering the A-site, the translocation to the P-site, and the deacylated tRNA exiting from the E-site are, in fact, rather simplified. Instead, it seems that the events are somewhat more complicated and follow the so-called hybrid states model. In this model, the ternary complex occupies first the T-site adjacent to the A-site. Then there is a state where the complex shares the T-site and the A-site before being in the A-site; likewise, there is a sharing of the P-site and the A-site before translocation to the P-site (Figure 10.15).

Having outlined the basic events in translocation, we will proceed now to examine in detail the mechanisms whereby these events take place. Obviously, for the events to occur, dynamic changes must take place. The ribosome is a

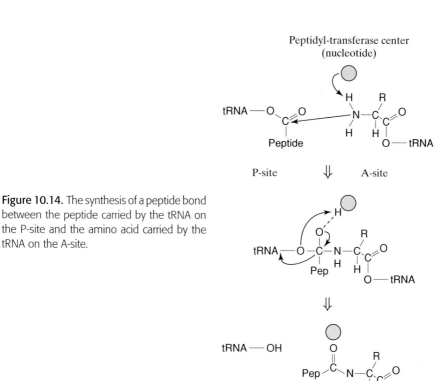

Figure 10.14. The synthesis of a peptide bond between the peptide carried by the tRNA on the P-site and the amino acid carried by the tRNA on the A-site.

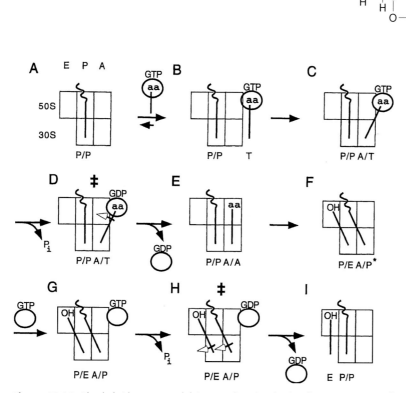

Figure 10.15. The hybrid-states model for translocation in the ribosome. H. F. Noller, Cell 92: 337–49 (1998). Reprinted with permission from Elsevier Science.

complex structure made up of proteins and rRNAs. It is conceivable that, during the events of translocation, specific interactions between the components of the ribosomes and the aatRNA ternary complex must be enforced. The availability of data at the 3-D level will be most helpful in understanding these mechanisms. Most of the structural studies have used prokaryotic factors and ribosomes. Therefore, we will concentrate our three-dimensional analysis on prokaryotic protein synthesis.

The Structure of Elongation Factors EF-Tu and EF-G

EF-Tu is a three-domain protein. Domain 1 (N-terminus; red to yellow in Figure 10.16) is the GTP-binding domain. In Figure 10.16, domain 2 is depicted as green/turquoise, and domain 3 is shown as blue. Depending on whether GTP or GDP is bound, the structure of EF-Tu undergoes conformational changes. The most important change is that, upon hydrolysis of GTP to GDP, domain 2 rotates about 90 degrees. The ET-Tu.GTP structure creates a cleft between

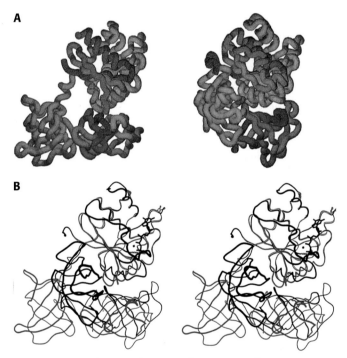

Figure 10.16. A: The three-dimensional structure of *Thermus aquaticus* EF-Tu. The GDP-bound structure is on the left, the GTP-bound structure is on the right. The color code is red to blue from the N-terminus to the C-terminus. Note the change in the conformation of domain 2 in regard to the rest of the molecule. In the GDP-bound structure, domain 2 has rotated about 90 degrees. **B:** This conformational change can be seen in stereo with the two structures superimposed. Black is the GTP-bound form, and red is the GDP-bound form. Note the tRNA-binding cleft that was created after GTP binding. The GDP bound in domain 1 is also shown. N. Kjeldgaard, Structure 1: 35–50 (1993). Reprinted with permission from Elsevier Science.

domain 1 and domain 2 and between domain 1 and domain 3. The stereo image in Figure 10.16B depicts this conformational change very clearly. This cleft contains amino acids that have been implicated in aatRNA binding, and these sequences are very conserved. Because EF-Tu interacts with many different tRNAs, it is expected that the tRNA-binding site should be conserved. Therefore, the conformational change that occurs when GTP binds exposes the aatRNA-binding site.

The three-dimensional structure of EF-G reveals something very interesting. The protein structure is very similar to the EF-Tu.GTP.tRNA ternary complex (Figure 10.17). This molecular mimicry provides important clues for the function of EF-G and its role in translocation. As mentioned earlier, after the peptide bond has been made in the A-site, EF-G binds near the A-site and promotes translocation of the tRNA to the P-site. It is conceivable, therefore, that, for EF-G to occupy the A-site, it must not only be of the correct size and shape but also be similar to EF-Tu.GTP.tRNA, which normally occupies the A-site.

Now that we have outlined the general events of translocation and the structures of the factors involved, let us view the process of translocation using the three-dimensional structures of the tRNAs, ribosome, and translocation factors (Figure 10.18).

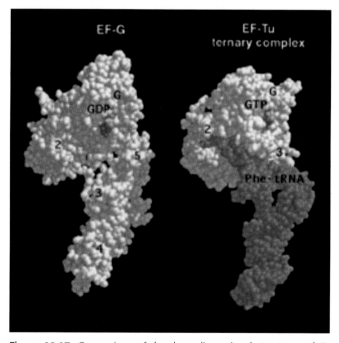

Figure 10.17. Comparison of the three-dimensional structures of *E. coli* EF-G (left) and EF-Tu.GTP.tRNA (right). Numbers indicate the domains of the proteins. Note how similar in size and shape these structures are. H. F. Noller, Cell 92: 337–49 (1998). Reprinted with permission from Elsevier Science.

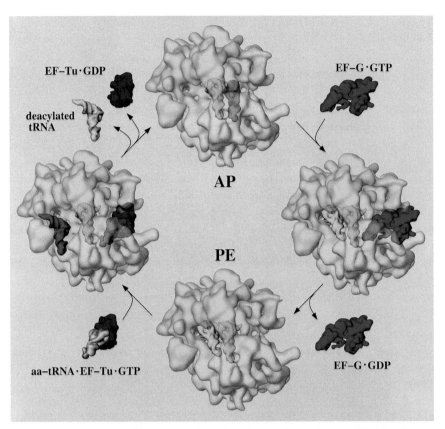

Figure 10.18. A 3-D view of the process of translocation. The large subunit is light blue, the small subunit is yellow, the tRNA in the A-site is pink, the tRNA in the P-site is green, the exiting tRNA is bright yellow and brown, EF-G.GTP is purple, the new tRNA to enter the A-site is gray, and EF-Tu.GDP or EF-Tu.GTP is red. In the top panel, both P- and A-sites are occupied by tRNAs (the lighter version of their color code indicates that these portions are behind (covered) the small subunit. Peptide bond formation is in the A-site (not shown). After the peptide bond is made, translocation is triggered by EF-G.GTP (purple), which bind at the vicinity of A-site (T/A site). After translocation, GTP is hydrolyzed to GDP, and EF-G.GDP leaves the ribosome. The A-site is now ready to be occupied by the new aatRNA.EF-Tu.GTP ternary complex. J. Frank, Amer. Scientist 86: 428–39 (1998). Courtesy of Dr. J. Frank.

Movement of tRNAs During Translocation

Let us now visualize the movement of the tRNAs during the process of translocation and elongation. In Figure 10.19, we can observe the positions of the tRNAs, bound on the 30S subunit, as can be seen from the 50S interface side.

Having visualized the positions of tRNAs on the 30S subunit, let us look at the same positions with a different 3-D model (electron density maps) (Figure 10.20A) and compare the view from the 50S interface with the view from the 30S interface (Figure 10.20B).

We can now proceed to view these events of translocation in the *E. coli* 70S ribosome. Figure 10.21 provides stunning stereo views of the positions and

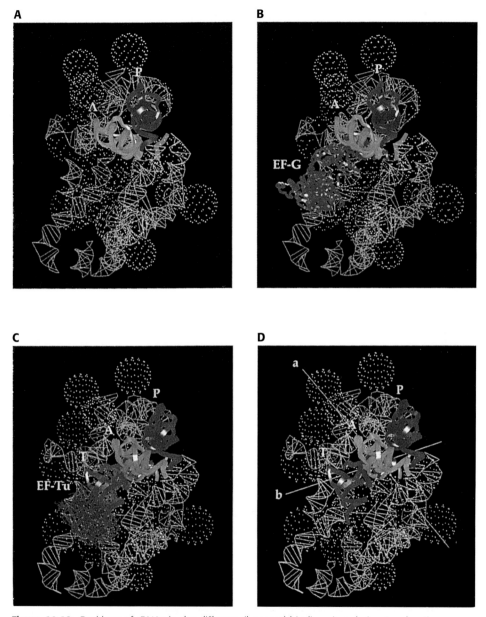

Figure 10.19. Positions of tRNAs in the different ribosomal binding sites during translocation. The tRNA is light blue on the A-site (AA in the hybrid state) and red on the P-site. EF-G is magenta, and EF-Tu is red. In A we can see the positions of tRNA in the A- and P-sites. In B the EF-G, with its domain 4, overlaps with the anticodon arm of the aatRNA. As already mentioned, this overlapping will trigger translocation. In C, we can see the same arrangement when the EF-Tu.GTP.tRNA ternary complex is in the T/A-site. The anticodon arm follows a similar path with domain 4 of EF-G, overlapping with the anticodon of the aatRNA. In D, the positions of tRNA in the T/A, A/A, and P/P states are visible. The planes of the tRNAs are related by a 60-degrees rotation along the a-axis (path of translocation) and 60-degrees rotation around the b-axis of the anticodon. H. F. Noller, Cell 92: 337–49 (1998). Reprinted with permission from Elsevier Science.

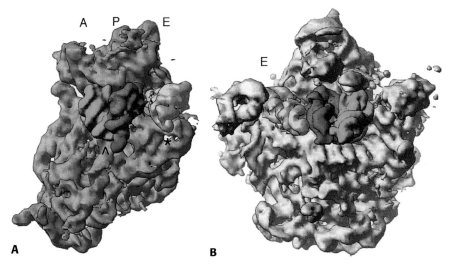

Figure 10.20. A: Three tRNAs (green at A-, blue at P-, and yellow at E-site) viewed on the small subunit interface (light blue). Note the 3'-CCA end of the A-site tRNA (^) and the E-site tRNA (*). On the E-site, the 3' end is buried at the base of the L1 stalk. **B:** Same tRNAs viewed on the 50S interface. Note that the 3' end of the A-site tRNA is by the L7/L12 region and that the E-site tRNA exits through the L1 stalk. Structures from *Thermus thermophilus.* H. Noller, Science 285: 2095–104 (1999). Reprinted with permission from American Association for the Advancement of Science.

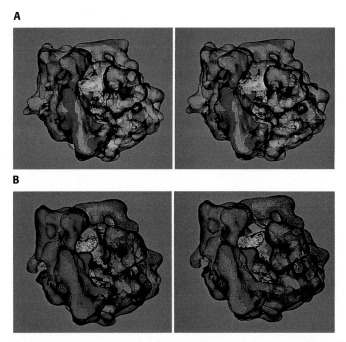

Figure 10.21. A: A stereo view of the tRNA on the A-site. Note interaction of the 3' end with the L7/L12 region. **B:** The tRNA on the P-site. Note the rotation and the interaction of the 3' end with the L7/L12 region.

C

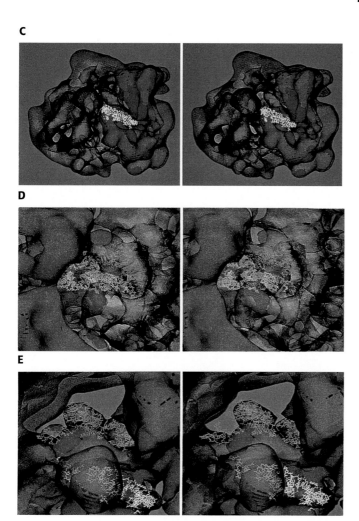

D

E

Figure 10.21(cont.) **C:** The tRNA on the E-site. Note that the ribosome has been turned around to visualize the exit site. The tRNA passes through the L1 region. **D** and **E:** Higher magnification of the pretranslocation (D) and the posttranslocation (E) stages. Both A- and P-site tRNAs are modeled, in D, while in E both P- and E-site tRNAs are modeled. The interaction between L7/L12 (brown) with the 3′ end of the tRNAs can be seen in D. The exiting tRNA passes underneath the L1 in E (that is why part of the tRNA, colored yellow, is underneath L1) as also was indicated in Figure 10.21C. Structures are from *E. coli* ribosome. M. van Heel, Cell 88: 19–28 (1997). Reprinted with permission from Elsevier Science.

translocation of tRNAs. The 50S subunit is colored gray with the L7/L12 and the L1 sides being brown. The tRNA is light blue on the A-site, green on the P-site, and white on the E-site.

Movement of EF-G and Conformational Changes of the Ribosome During Translocation

The reader should already be anticipating that translocation and elongation affect the conformation of the other players, such as the ribosome and the EF-G. In the following studies, investigators were able to catch snapshots of the ribosome with EF-G before and after translocation by using thiostrepton, which considerably slows the process of translocation. By monitoring the incubation time, they were able to receive ribosome at the pre- or post-translocation stage. The structural studies revealed that the ribosome and EF-G undergo large changes in their conformation. In Figure 10.22, we can see the interactions of the EF-G with the ribosome and those conformational changes that

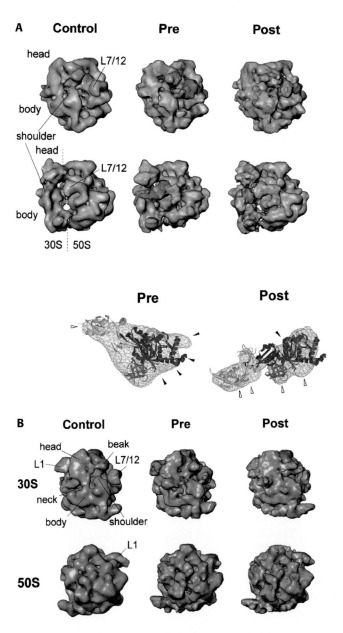

occur during translocation. All these changes coordinate with conformational changes of the ribosome. Major changes can be seen at the neck that connects the head with the body and at the connection between the head and the beak. The large subunit appears mostly unchanged (Figure 10.22B).

INTERACTIONS BETWEEN THE 30S AND 50S SUBUNITS

Before we examine the interactions between components of the ribosome that eventually will help us understand the interactions with tRNAs at the decoding center, the reader should become familiarized with the 3-D structure of the ribosome. In previous sections, we examined 3-D images of whole ribosomes, but it is important to know the anatomy of the 30S and 50S subunits individually. In Figure 10.23, we will examine stereo views of the interface between the ribosomal subunits. The top panel represents the 50S subunit (gray), and the 30S subunit (blue) is below it. Regions of contact between the two subunits are traced with black lines, and an illustration is presented for clarity. Since the interface between the two subunits is like the pages of an open book, the contact areas are mirror images. Note on the 30S subunit the so-called penultimate stem, or helix 44, which contains three contact regions with the 50S subunit – B2a, B3, and B5. Other important areas are the B2b, which contains the 790 helix (or helix 24); B7, which contains the 690 helix (or helix 23); and the B2c, which contains the 900 loop (which is flanked by the switch helix, or helix 27). The position of these areas in the 16S rRNA will become apparent later. Contacts B2a, B2c, B3, B5, and B7 are RNA-RNA interactions while contacts B1a, B1b, B4 and B6 are most likely RNA-protein interactions (Figure 10.23). All these areas are extremely important if the reader is to visualize the molecular interactions where decoding is occurring.

Figure 10.22. **A:** Arrangement and conformation of the *E. coli* EF-G on the *E. coli* ribosome during translocation. The left panel shows the control ribosome (no EF-G bound). The middle panel shows the ribosome with EF-G bound at the pre-translocation stage. The right panel shows the ribosome and EF-G at the post-translocation stage. The first and the second rows show two different views of the ribosomes rotated 30 degrees. The 30S subunit is on the left, the 50S is on the right, and EF-G is blue. The third row contains 3-D images of EF-G at the pre-translocation and the post-translocation stages. We can see that at the pre-translocation complex EF-G interacts mainly with the L7/L12 region of the 50S subunit. In fact, the interaction (closed arrowheads) is restricted mainly with domain 1 (purple in the 3-D image in the third row). Domain 4 (yellow) interacts with the 30S subunit (open arrowhead). However, at the post-translocation complex, there are major changes in the conformation and the interactions of EF-G with the ribosome. The major interaction between EF-G and L7/L12 has been lost, and now the interactions with the body and head of the 30S are quite extensive (open arrowheads). Domain 5 (red) is displaced toward domain 2 (blue). This results in a 45-degree rotation of domain 4 (yellow). **B:** Conformational changes in the ribosome. The upper row is viewed from the solvent side of the 30S, and the lower row is seen from the solvent site of the 50S subunit. Note the changes in the 30S subunit. W. Wintermeyer, Cell 100: 301–9 (2000). Reprinted with permission from Elsevier Science.

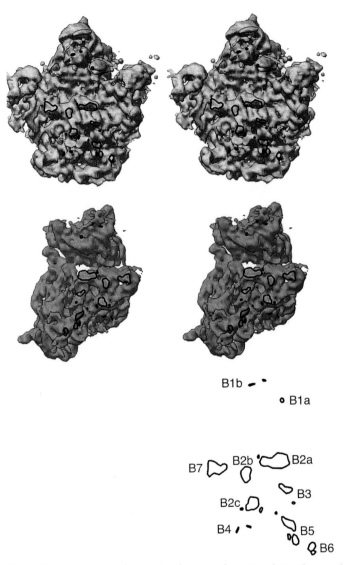

Figure 10.23. Contacts in the interface between the 30S and 50S ribosomal subunits of *Thermus thermophilus* (see text for explanations). H. Noller, Science 285: 2095–104 (1999). Reprinted with permission from American Association for the Advancement of Science.

Having identified the areas of interactions between the ribosomal subunits, let us see the structure of the 16S rRNA so that we can recognize these areas and examine their 3-D structure and placement in the 30S ribosomal subunit. Figure 10.24A shows the 2-D structure of the 16S rRNA with the different domains colored specifically and the helices numbered. In Figure 10.24B, we can see how the 16S rRNA domains fold in three dimensions. The composite of Figure 10.24B should familiarize the reader with the 3-D structure of the

A

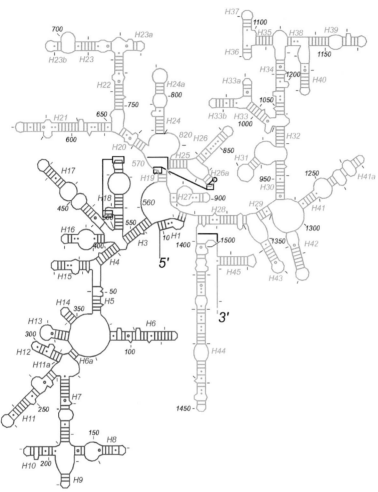

Figure 10.24. A: The 2-D structure of *Thermus thermophilus* 16S rRNA with its domains identified with different color codes (red, 5′ domain; green, central domain; brown, 3′ major domain; cyan, 3′ minor domain, which contains helix 44).

30S subunit and the location of the important areas of contact with the 50S subunit. In this regard, note that the 16S rRNA folds in such a way so that helix 27 interacts with helix 44 and that other important regions fold in such a way as to be close to the catalytic center (see also later).

Interactions Between tRNA and rRNAs

So far we have seen the general mechanisms of translocation, the position of tRNAs during this process, and the 3-D structure of the 30S subunit. As has been alluded to previously, the tRNAs interact specifically with both

B

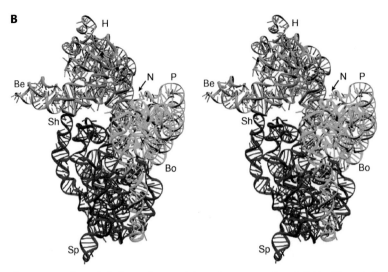

Figure 10.24 (cont.) B: Stereo image of the 3-D structure of 16S rRNA with each domain colored with the same color code as in A. H, head; N, neck; P, platform; Sh, shoulder; Sp, spur; Bo, body; Be, beak. V. Ramakrishnan, Nature 407: 327–39 (2000). Reprinted with permission from Nature, Macmillan Magazines Ltd.

subunits of the ribosome. The reader must be already familiar with the fact that the anticodon region of the tRNA interacts with the 30S subunit, while the 3′ acceptor end interacts with the 50S subunit. Also, we know that the prokaryotic 30S subunit is composed of the 16S rRNA and the 50S of 23S rRNA and 5S rRNA. In the following discussion, we will examine the interactions between the tRNAs in the A- and P-sites with the different rRNAs.

Interactions Between tRNA and 16S rRNA

Such interactions are depicted in the stereo image presented in Figure 10.25. These interactions have been revealed by chemical methods and modeled according to the positions of tRNA in the different sites. The interaction is shown by a colored cloud, each cloud representing the interaction of a region of 16S rRNA (identified by a nucleotide number). The stronger the interaction, the smaller the cloud. Thus, we can observe that 16S rRNA interacts strongly with the anticodon stem-loop. The regions of interaction belong to the 690 helix (helix 23; B7) and 790 helix (helix 24; B2b). Weaker interactions can be seen with regions around positions 1300 and 1338. Loop 1338 is located in the head of the 30S subunit lining the inside of the cleft. Also note that there are no contacts with the tRNA on the A-site.

Let us now see this interaction between the A- and P-site tRNA with the 16S rRNA at the 3-D level (Figure 10.26). Three regions (designated as fingers a, b, and c) of the 30S subunit create a clamp that holds the anticodon stem loop (ASL) in the P-site. These regions are shown as nets in Figure 10.26A. Finger d

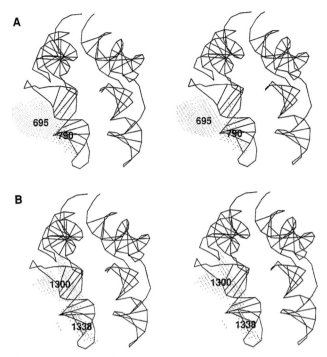

Figure 10.25. Interactions of tRNAs with the 16S rRNA. The tRNA on the left on the P-site is and on the right on the A-site. The contacts by nucleotides of the rRNAs are indicated with colored pixel clouds and are numbered. The strong and weak interactions are shown in A and B, respectively. H. Noller, Science 278: 1093–8 (1997). Reprinted with permission from American Association for the Advancement of Science.

contacts the backbone of the anticodon loop near the wobble nucleotide. Fingers d and e grip the codon-anticodon duplex from opposite directions. Finger f merges with the codon-anticodon helix, indicating a possible stacking interaction with the wobble pair. Finger f most likely contains nucleotides from the 1400 region of the 16S rRNA very close to the penultimate stem (helix 44).

Interactions Between tRNA and 23S rRNA

Before we examine interactions between tRNA and 23S rRNA, we should examine the structure of 23S rRNA and its place in the large subunit. As in the case of 16S rRNA, 23S rRNA will help us place these interactions in relation to the process of translocation and protein synthesis. The secondary structure of 23S rRNA is, of course, very complex; consequently, it is very difficult to visualize the interactions between proteins and tRNA as revealed by biochemical data. However, we will attempt to follow a sequence that could be very helpful. The secondary structure of the 23S rRNA reveals that it is composed of six domains. This can be seen in Figure 10.27, where the helices are numbered. Domain I is from helix 1–25, domain IIa is from helix 26–36 and helix 46,

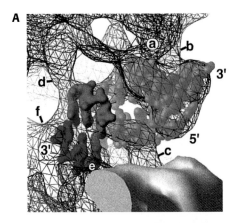

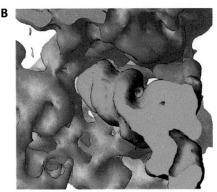

Figure 10.26. A: Interactions between the anticodon stem loop (light blue) with regions a through e of the *Thermus thermophilus* 16S rRNA (purple net). The anticodon is dark blue, and the P-site codon is red. The large subunit is gray. This view is from the A-site. **B:** The A-site binding pocket on the 30S subunit (purple) with the A-site tRNA (green). The P-site ASL is blue. H. Noller, Science 285: 2095–104 (1999). Reprinted with permission from American Association for the Advancement of Science.

domain IIb is from helix 37–45, domain III is from helix 47–60, domain IV is from helix 61–71, domain V is from helix 72–93, and domain VI is from helix 94–101. In this two-dimensional map of the 23S, rRNA we can note a few very important places. The 23S rRNA has a conserved core, which is presented with black bars. This core is conserved between prokaryotes and eukaryotes. The interaction with L11 and EF-G is with helix 43 and 44 of domain IIb. The EF-G and EF-Tu interact with the loop of helix 95 (where the sarcin-ricin loop sequences also are; see later). The central loop of domain V, the peptidyl ring, is enclosed by helices 73, 74, 89, 90, and 93. Interactions of 5S rRNA are with L25 and helices 83–85 (see also later).

Let us now put the two-dimensional map of 23S rRNA into a 3-D perspective as it fits in the large ribosomal subunit. Using a color code for each domain and representing each helix as a numbered cylinder, we can see their location in the

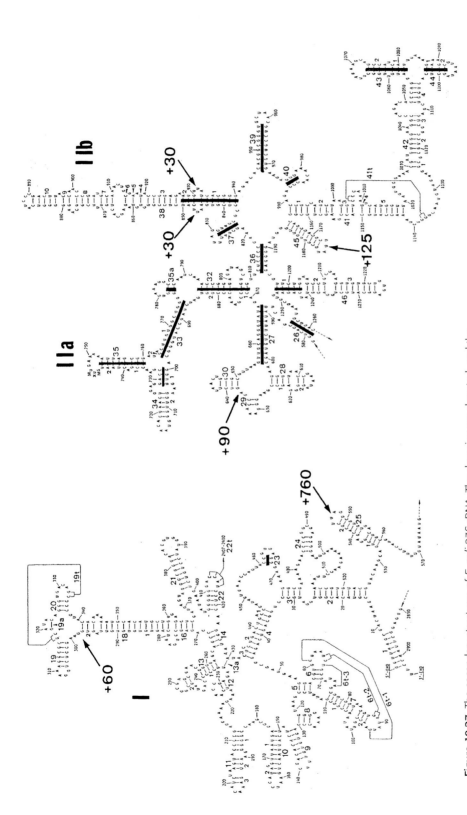

Figure 10.27. The secondary structures of *E. coli* 23S rRNA. The domains are denoted, and the conserved core is shown with the black bars. Arrows with numbers indicate the insertions and their sizes in nucleotides found in rat 28S rRNA. R. Brimacombe, J. Mol. Biol. 298: 35–59 (2000). Reprinted with permission from Academic Press Ltd.

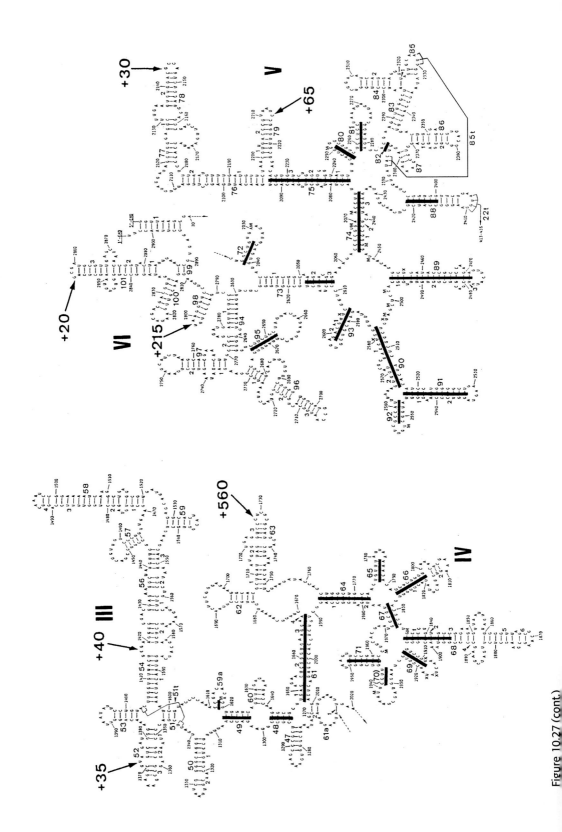

Figure 10.27 (cont.)

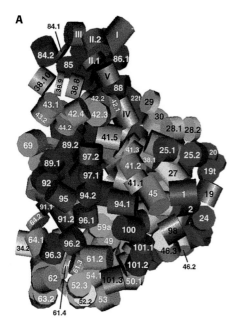

A

Figure 10.28. A: The 3-D arrangement of the double helical elements of the *E. coli* 23S rRNA and 5S rRNA as seen from the A-site. Domain I is red, IIa is yellow, IIb is orange, III is green, IV is light blue, V is magenta, and VI is dark blue. 5S rRNA is brown. **B:** The arrangement of the 23S rRNA helical domains as seen from the interface. **C:** Same as in B, but with only the core helical elements shown. R. Brimacombe, J. Mol. Biol. 298: 35–59 (2000). Reprinted with permission from Academic Press Ltd.

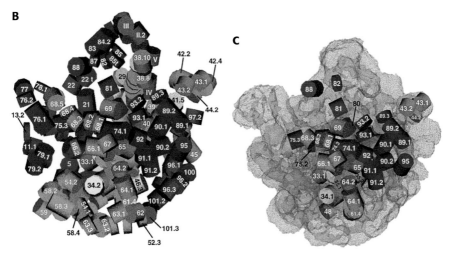

large subunit. In Figure 10.28A, the view is from the L7/12 site; in Figure 10.28B it is from the interface, and in Figure 10.28C it is the same view as in Figure 10.28B, but only for the conserved core. Obviously, domain I, which has no conserved core elements, comprises the solvent area of the large subunit. Domains IV and V occupy the interface area where most of the interactions with the A- and P-site tRNA occur and the peptide bond is made (see also later).

Having examined the 3-D structural features of 23S rRNA, let us now consider its interactions with tRNA in more detail. In Figures 10.29A and 10.29B, we can observe interactions of the 23S rRNA with the A- and P-site tRNA.

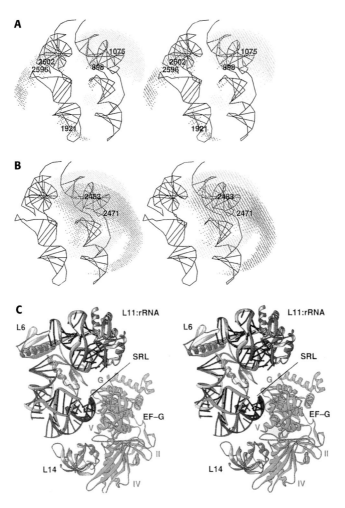

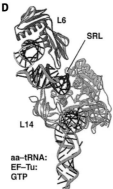

Figure 10.29. A and **B:** Interactions of tRNA with the *E. coli* 23S rRNA. See legend of Figure 10.25 for more details. H. Noller, Science 278: 1093–8 (1997). **C** and **D:** The *Haloarcula marismortui* GTPase center of the large subunit and interactions near the A-site with EF-G:GTP and aa-tRNA:EF-Tu:GTP respectively. The blue and yellow RNAs are most likely helices 96 and 97 of the 23S rRNA. T. A. Steitz, Nature 400: 841–7 (1999). A and B reprinted with permission from American Association for the Advancement of Science. C and D reprinted by permission from Nature, Macmillan Magazines Ltd.

In Figure 10.29A, strong interactions are shown; weak interactions are shown in Figure 10.29B. A strong interaction occurs between the 2600 region and the P-site tRNA. This region of the 23S rRNA has been placed near the peptidyl transferase center. Also, a region (1060–1100), which is known to be associated with L11 and interact with elongation factor G, interacts with the A-site tRNA. Other regions of interaction include those sites of 23S rRNA that interact with antibiotics that inhibit translation (900) and the central loop of domain V (2455–2496), which is placed in the vicinity of the tRNA acceptor ends. The 3-D structures at the A-site showing interactions between EF-G:GTP or aatRNA:EF-Tu:GTP and 23S ribosomal RNA and proteins can be seen in Figures 10.29C and 10.29D, respectively. We can see that the sarcin-ricin loop sequences (SRL; helix 95; red); the L11-associated 1055- to 1080-nucleotide region of 23S rRNA (dark yellow); L6, L11, and L14 proteins (gray); EG-Tu and EF-G (light green); and aatRNA (dark green). This grouping basically comprises the GTPase center or factor-binding site, which, in addition to the aforementioned RNAs and proteins, contains the L7/L12 stalk. The importance of this region has been demonstrated with experiments involving inhibition of elongation factor function due to antibiotic treatment. Antibiotics such as those belonging to the thiostrepton family, most likely bind in a narrow cleft between the RNA and the N-terminus domain of L11. The central position of SRL suggests that it might stimulate GTPase activity of factors like EF-Tu and EF-G when they are bound to the ribosome.

Let us now see more of these interactions in stereo. Figure 10.30A shows positions of A- and P-site tRNAs viewed from the L7/12 side. In the 3-D structure, the deep cleft where the peptide bond occurs is obvious in the A-site. The reader should be accustomed to seeing the domelike structure that represents the place where the A- and P-site tRNA and factors bind. We can see the interactions of the tRNAs with relevant regions of 23S rRNA (blue backbone tubes with helices numbered). The colored spheres denote nucleotides that are contacted. Of interest is the interaction of both tRNAs with nucleotides 2439 and 2451 of helix 74 and the following loop (nucleotides are shown as magenta spheres). In Figure 10.30B, we can see the P-site tRNA (red backbone tube) interacting with the peptidyl transferase center. Colored spheres denote the cross-linked nucleotides, and X denotes the exiting tunnel for the peptide. We can see the exiting channel, which spans the large subunit in Figure 10.30C. The P-site tRNA is again a red backbone tube, and the path of the tunnel is shown as a blue arrow. The spheres denote interactions of nucleotides with different size peptides. Red is nucleotide 2609 cross-linked to a peptide with three amino acids, orange is nucleotide 1781 (four residues), green is nucleotide 750 (six residues), turquoise is nucleotide 1614 (12 to 25 residues), and magenta is nucleotide 91 (30 residues). These studies demonstrate very elegantly how the growing peptide interacts with components of the large subunit as it exits through the tunnel.

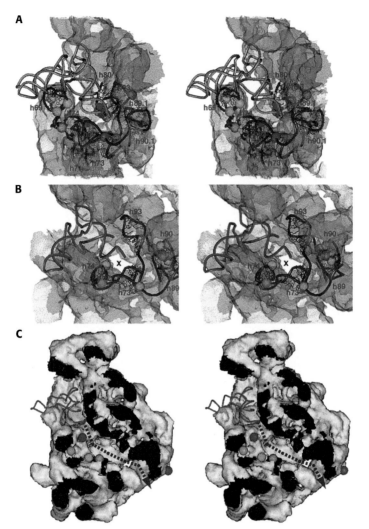

Figure 10.30. A: Interactions of A-site tRNA (light blue) and P-site tRNA (green) with the *E. coli* 23S rRNA as characterized by cross-linking biochemical data and fitting in the model. The view is from the L7/12 side. **B:** Interactions of the P-site tRNA (red) with the peptidyl center. Elements of 23S rRNA are represented by blue backbone tubes, and X is the peptide exit tunnel. The view is from the interface side of the 50S subunit. **C:** The peptide exit tunnel is revealed by a section of the 50S subunit and is seen from the L7/12 side. R. Brimacombe, J. Mol. Biol. 298: 35–59 (2000).

Interactions Between tRNA and 5S rRNA

The 5S rRNA is part of the central protuberance (CP) (see Figure 10.28). However, its interactions with tRNA have been mapped mostly above and to the left of P-site tRNA, which is consistent with its place in the CP. 5S rRNA is associated with L25 (Figure 10.31) and interacts with helices 83–85 and 89 of domain V and helix 38 of domain IIb.

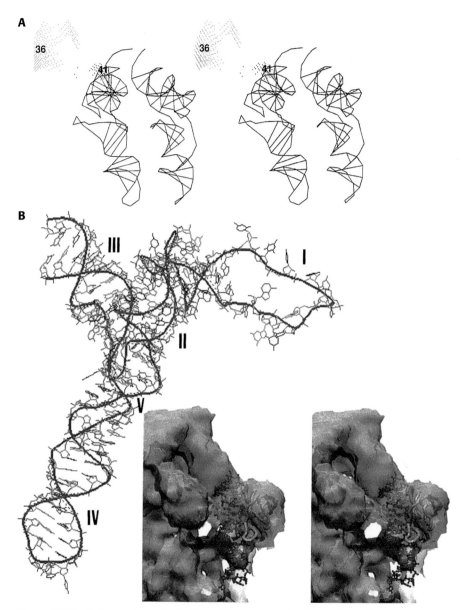

Figure 10.31. **A:** Interactions of 5S rRNA with tRNA. For details see legend of Figure 10.25. H. Noller, Science 278: 1093–8 (1997). A reprinted with permission from American Association for the Advancement of Science. **B:** A 3-D model of *E. coli* 5S rRNA with its helices denoted I–V. In the insert, its association with L25 can be seen in stereo. B. R. Brimacombe, J. Mol. Biol. 298: 35–59 (2000). Reprinted with permission from Academic Press Ltd.

If we look at Figures 10.25, 10.29, and 10.31 again, some interesting patterns emerge. First, there are no interactions of ribosomal RNA in the space between the A-site and P-site. This barrenness probably provides undisturbed movement of tRNAs from the A-site to the P-site. Also, the A-site clouds are more diffuse than those in the P-site. One explanation for this distribution is that the ASLs have greater freedom of movement in the A-site, due to weaker affinity in the A-site for tRNA. Also this distribution could be due to the required significant movement of the tRNA from the A/T to A/A state, while maintaining codon-anticodon interactions.

THE DECODING CENTER OF THE RIBOSOME

So far we have seen the interactions of tRNAs from the A-, P-, and E-sites with the ribosomal RNAs. The 30S A- and P-sites (interacting with tRNAs and mRNA) are mostly composed of rRNA, and the E-site is mostly composed of proteins, especially S7 and S11 (Figure 10.32A). Having examined the molecular anatomy of the interactions in the ribosome during translocation let us now take another magnified look at the decoding center. In the heart of the decoding center of the 16S rRNA are the regions between 1404–1412 and 1488–1497. These regions are on the penultimate stem (helix 44) near the 5' and 3' ends (Figure 10.32B). This region also interacts with the switch helix (helix 27) (Fig. 10.32C,D; see also Figure 10.24).

A1492 and A1493 are universally conserved. They are imperative for viability in *E. coli* and are footprinted by the A-site codon and tRNA. In Figure 10.32D, these two bases are shown away from the codon-anticodon helix. However, in the flipped-out conformation, these two bases would be able to interact with the minor groove of the codon-anticodon helix (Figure 10.33). Remember that IF-1 also interacts with these two nucleotides (Figure 10.9). It is proposed that the A1492 and A1493 adenine bases interact via hydrogen bonding simultaneously with the 2'-OH groups on both sides of the codon-anticodon helix. The two adenines can monitor the minor groove of three consecutive base pairs. Thus, since the 2'-OH groups have a distance characteristic of base-pair geometry, the hydrogen bonding with both strands would be sensitive to distortions resulting from mispairing. This can monitor the correct shape of the codon-anticodon helix and allow only the cognate tRNA to interact with each codon and not other near-cognate tRNAs (Figure 10.33).

Let us elaborate some more about the importance of these structures and interactions in the process of translocation. It has been suggested that the tRNA-bound ribosome is at a locked state. EF-G catalyzes translocation by unlocking this trap in a GTP-dependent manner. Thus, unlocking should involve disruption of tRNA-ribosome interactions as well as rearrangement of intramolecular interactions within the ribosome. These alterations can cause the ribosome to tighten or loosen its grip on tRNA. Indeed, there is enough evidence indicating that such alterations do take place. Initially, it was found that mutations

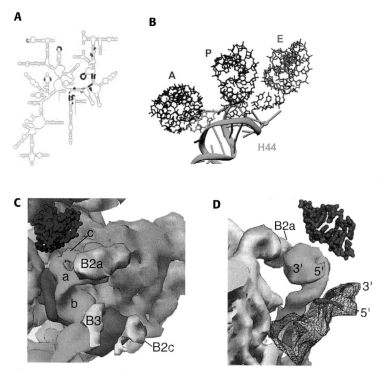

Figure 10.32. **A:** Secondary structure of 16S rRNA indicating the RNA elements involved in interactions in the A-site (magenta), P-site (red), and E-site (yellow). **B:** Model showing the A-, P-, and E-site tRNAs (magenta, red, and yellow, respectively) interacting with codons at each site (light blue, dark blue, and purple). Note the close association with helix 44. The red bars indicate nucleotides 1492 and 1493. V. Ramakrishnan, Nature 407: 340–8 (2000). **C:** Another view of the decoding center in the *Thermus thermophilus* 16S rRNA. The small subunit is purple; the large subunit is gray (the areas B2a, B3 and B2c interact with the 30S subunit). A-site tRNA is red, mRNA is dark purple, helix 44 is green, and the switch helix is blue: a is bases 1492 and 1493 (far away from the A-site tRNA or mRNA, but near the switch helix), b is the location of nucleotides 909 (from switch helix), 1413, and 1487. These three nucleotides are called class III nucleotides, and they interact with tRNA or 50S subunit. c is nucleotide 1494. **D:** A different view of the center showing the switch helix in ribbon model. H. Noller, Science 285: 2095–104 (1999). A and B reprinted by permission from Nature, Macmillan Magazines Ltd. C and D reprinted with permission from American Association for the Advancement of Science.

in alleles of S12 protein that confer streptomycin resistance lead to a hyper-accurate (restrictive) phenotype (remember that antibiotics can interfere with elongation factor function). On the other hand, mutation in S4 and S5 that suppress antibiotic resistance lead to increased translational errors (*ram*, or ribosomal ambiguity phenotypes). Obviously, mutant proteins can exert their effects by altering the 3-D conformation of 16S rRNA. In addition to mutated proteins, mutations in 16S rRNA can produce streptomycin resistant and *ram* phenotypes. In particular, in helix 27 (switch helix) there can be two alternative pairings, one producing restrictive phenotypes and the other producing *ram* phenotypes (Figure 10.34).

A

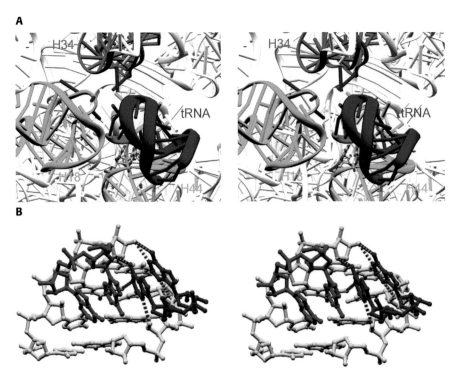

B

Figure 10.33. Role of A1492 and A1493 in decoding in *Thermus thermophilus*. **A:** Stereo image of the A-site. The anticodon stem loop is red, helix 44 is light blue, helix 34 is dark blue, and helix 18 is green. The ball-and-stick models are for A1492 and A1493 in the flipped-out conformation and the red bars indicate some other highly conserved residues that might help A1492 and A1493 recognize the codon-anticodon helix. **B:** Stereo image of the hydrogen bonding between A1492 and A1493 (red) and the codon-anticodon helix. Note that the simultaneous hydrogen bonding measure 2′OH-2′OH distances for three successive base pairs. Blue bases are other conserved bases that might help A1492 and A1493. V. Ramakrishnan, Nature 407: 340–8 (2000). Reprinted by permission from Nature, Macmillan Magazines Ltd.

Recall that this helix is at the heart of the decoding center and that the 900 loop contains nucleotide 909, which interacts with the penultimate stem (helix 44) (Figure 10.34). The *ram* phenotypes result in an S-turn for the 5′ strand (in fact, this is the case in the crystal structure of Figure 10.32). It

Figure 10.34. **A:** Movement of CUC of helix 27 (switch) by 3 nucleotides, which results in the restrictive or *ram* phenotypes. **B:** Helix 27 with the triplet (885–887) colored red. CUC is colored blue, and the S-turn of the *ram* conformation is green. **C:** Interactions of helix 27 with helix 44, with residues whose chemical reactivity is affected in the restrictive S12 strains labeled. **D:** The environment of helix 44, helix 27, and S12. Red balls indicate the sites of restrictive mutations in S12, and red sticks are A1492 and A1493 (stereo image). **E:** The environment of helix 18, S4, and S5. Red highlights the location of *ram* mutations in S4 and S5 (stereo image). V. Ramakrishnan, Nature 407: 340–8 (2000). B, C, D, and E reprinted by permission from Nature, Macmillan Magazines Ltd.

A

Restrictive

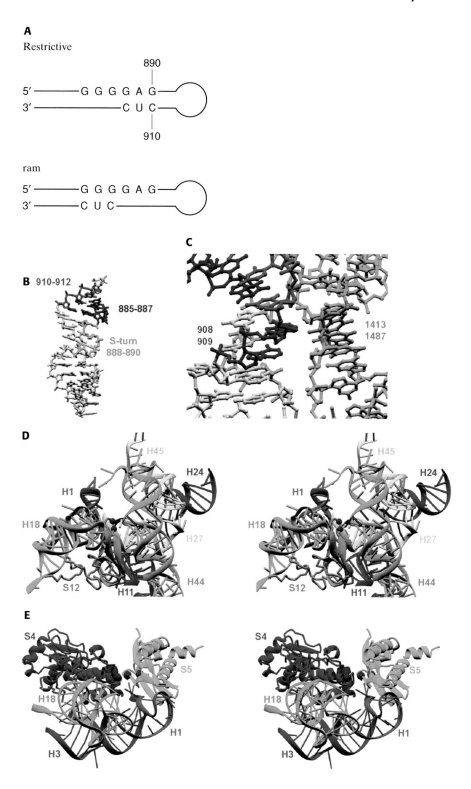

B

C

D

E

is interesting to note that the switch, in other words the movement of CUC by exactly 3 nucleotides, does resemble the process of translocation. It is very possible, therefore, that the intramolecular conformational changes within the ribosome lead to the process of unlocking and translocation.

THE PEPTIDE BOND CENTER

Let us look again at the large subunit interface of the *Haloarcula maris-mortui* solved at 2.4-Å resolution (Figure 10.35). Sequences that are >95% conserved across the three phylogenetic kingdoms are shown in red, and expansion sequences are shown in green. We can see clearly that much of the conserved sequences are located in the peptidyl transfer center. In Figure 10.35B, we can see a spacefill model of the large subunit. The RNA bases are white,

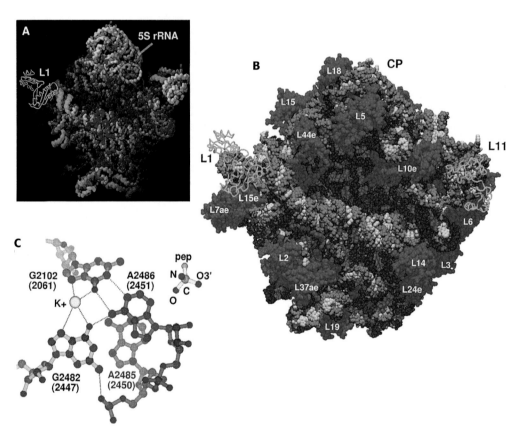

Figure 10.35. A and **B:** Model of the large subunit from *H. marismortui* indicating the conserved rRNA sequences (red in A) and the RNA-dominated region where the peptide bond center is located (white and brown). **C:** The catalytic apparatus of the peptidyl transferase active site. Hydrogen bonds are represented by dashed lines. The buried phosphate that could stabilize the interaction between G2482 and A 2486 is also shown. T. A. Steitz, Science 289: 905–30 (2000). Reprinted with permission from American Association for the Advancement of Science.

and the sugar-phosphate backbones are brown. The proteins are blue, and the CCA sequence, which is involved in peptide bond, is red. It is obvious from this model that the center where tRNAs are interacting with 23S RNA is occupied exclusively by RNA. There are no protein side-chain atoms closer than 18 Å to the peptide bond that is synthesized. Therefore, the catalytic apparatus of the peptidyl transferase active site is made of RNA. In fact, the four bases that are creating the catalytic site in *H. marismortui* are G2102 (2061), G2482 (2447), A2485 (2450), and A2486 (2451) (the corresponding bases in *E. coli* are in parenthesis). The arrangement of the catalytic site is shown in Figure 10.35C. Compare Figure 10.35C with Figures 10.27 and 10.30. Obviously, the ribosome is a ribozyme and catalyzes peptide synthesis by nucleophilic attack (see Figure 10.11). This catalysis is performed by base A2486 (2451 in *E. coli*). However, studies where A2451 or G2247 were mutated showed that catalysis was not affected. These studies might indicate that the ribosome promotes the formation of peptide bonds without chemical catalysis by positioning the reaction components in a configuration that will favor spontaneous reaction.

TERMINATION OF PROTEIN SYNTHESIS

Protein synthesis is terminated when an in-frame stop codon (UAG, UAA, or UGA) occupies the A-site. This process is mediated by the so-called release factors (RFs) that recognize the stop codons and promote the hydrolysis of the ester bond that links the polypeptide to the P-site tRNA. In prokaryotes, two similar factors, RF1 and RF2, are involved in this function; an unrelated GTP-binding protein, RF3, does not recognize a stop codon but stimulates RF1 and RF2 activity. Both RF1 and RF2 recognize UAA, but UAG is decoded by RF1 and UGA by RF2. In eukaryotes, eRF1 is the counterpart of RF1 and RF2 and recognizes all stop codons, while eRF3 is the equivalent of RF3. The molecular mechanisms of the stop codon decoding are largely unknown, but recent studies on the 3-D structure of human eRF1 have provided very crucial information about the process of termination.

The Three-Dimensional Structure of Human eRF1

Perhaps the structure was not a big surprise in that it very much looks like a tRNA molecule. It is conceivable that, if a protein recognizes a stop codon at the A-site, it has to mimic a tRNA in order to occupy that site. Indeed, the protein eRF1 is folded in such a way as to resemble tRNA in shape and size. eRF1 has three domains, with domain 1 corresponding to the anticodon stem and loop, domain 2 corresponding to the aminoacyl stem, and domain 3 corresponding to the T-stem (Figure 10.36).

Some of the important structural features in eRF1 should be noted. They play an important role in its function. First, the sequence GGQ, which is found in the tip of domain 2 is very conserved and corresponds to the CAA-3′ end

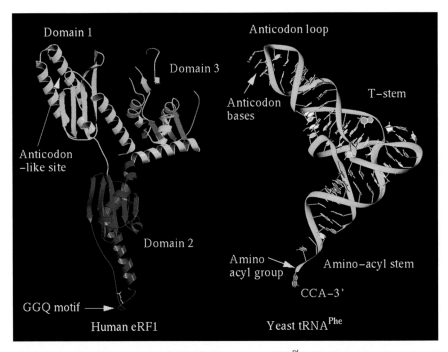

Figure 10.36. The 3-D structures of eRF1 (left) and yeast tRNA^{Phe} (right) revealing the molecular mimicry. D. Barford, Cell 100: 311–21 (2000). Reprinted with permission from Elsevier Science.

of the tRNA that binds the amino acid. It is conceivable that the conserved glutamine coordinates hydrolysis of the peptidyl-tRNA bond in the P-site by a water molecule (Figure 10.37).

Mechanism of Stop Codon Recognition

Studies with RF1 and RF2 have thrown light on the mechanism of stop codon recognition. As mentioned earlier, both factors recognize UAA, but RF1 is specific for UAG and RF2 is specific for UGA. How is this specificity achieved? RF1 has at its anticodon-like site the sequence Pro-Ala-Thr, while RF2 has Ser-Pro-Phe. With molecular biology techniques, it was found that the RF1 proline is a second-base (A) discriminator and that threonine is a

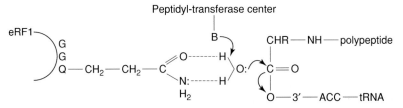

Figure 10.37. Proposed function of eRF1 in hydrolysis of the peptidyl-tRNA bond. Compare with Figure 10.14.

Figure 10.38. The *Thermotoga maritima* RRF superimposed with yeast tRNA^Phe. Note the almost perfect alignment between the two molecules. The only difference is that there is no overlap with the amino acid-binding 3′ end. A. Liljas, Science 286: 2349–52 (1999). Reprinted with permission from American Association for the Advancement of Science.

third-base (G or A) discriminator. Likewise, the RF2 serine is a second-base (G or A) discriminator, and phenylalanine is a third-base (A) discriminator. In other words, we have an equivalent tripeptide anticodon that is able to decipher stop codons. It is very interesting to note that the structural molecular mimicry between release factors and tRNA is extended to function as well, where there is a peptide equivalent to anticodons.

The Ribosome Recycling Factor

After termination and release of RF1 and RF2 from the ribosome by the help of RF3, the postermination complex consists of the 70S ribosome with a bound mRNA, an empty A-site, and a deacylated tRNA on the P-site. The disassembly of the ribosome is mediated by the ribosome-recycling factor (RRF) and EF-G (under GTP hydrolysis). RRF is very important for this event; without it, the ribosome cannot dissociate from the mRNA and continues protein synthesis downstream of the termination codon. The 3-D structure of the prokaryotic RRF is another example of molecular mimicry. As expected, RRF looks like a tRNA, which makes sense because this factor must occupy the A-site to trigger disassembly (Figure 10.38).

HOW DO ANTIBIOTICS WORK? A LESSON FROM THE 3-D STRUCTURE OF THE RIBOSOME

Antibiotics can kill bacteria by attacking many different targets in the cell. One of the favorite targets of antibiotics in the ribosome because we know that antibiotics do inhibit and interfere with protein synthesis. Consequently, it was suspected that antibiotics are bound onto the ribosome in places that are important for peptide synthesis, such as the A-site. With the determination of the 3-D structure of the ribosome, the mechanism whereby antibiotics bind and interfere with protein synthesis has become apparent. Let us, for example, consider the case of paromomycin. This antibiotic is a member of the aminoglycoside family of antibiotics and increases the error rate and initial binding affinity

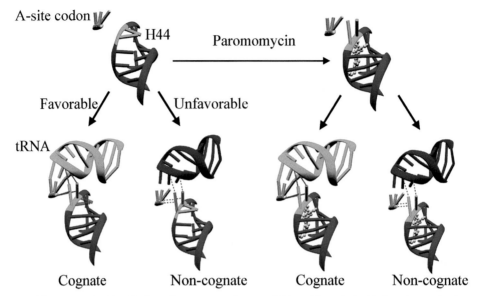

Figure 10.39. The binding of paromomycin onto H44 and its role in H44 codon-anticodon interactions. Courtesy of Dr. V. Ramakrishnan.

of tRNA. The 3-D structure of bound paromomycin on the 30S ribosomal subunit has been solved and provides clues relating to how it affects tRNA binding and errors. Paromomycin binds in the major groove of helix 44, and this binding allows A1492 and A1493 to flip out. Remember that this kind of conformational change in H44 is necessary for interaction with the codon and anticodon during decoding. As discussed earlier, the flipped A1492 and A1493 form hydrogen bonds with the codon-anticodon helix and help in the decoding. Obviously, the flipping-out of these bases requires some energy, and this cost is compensated by the favorable interactions with tRNA. Binding of paromomycin to H44 results in the flipping-out and reduces the cost of cognate and noncognate tRNA, thus increasing tRNA affinity for the A-site and presumably increasing errors in translation (Figure 10.39).

CHAPTER ELEVEN

The Birth and Death of Proteins

——— – – –

PRIMER As the polypeptide is born and exits from the ribosome, it is bound by factors that will enable it to pass through the membranes of the endoplasmic reticulum. This is possible because all secreted proteins contain a signal in the very N-terminus, the signal peptide. This is recognized by the signal recognition particle that brings it to the receptor in the membrane and facilitates the passing. Then, the polypeptide has to fold into its unique 3-D structure. We already know how important this is. The function of a protein depends on its 3-D structure. The folding must be correct; otherwise, the protein will not be allowed to perform its tasks. Folding involves many factors, called chaperones. The 3-D structures of chaperones also provide stunning images of the mechanism of folding. Alas, nothing lives forever! And this rule applies to proteins as well. In fact, it is central to regulation that the protein will die after some time. If a protein does not die, it will accumulate and induce diseases. Eventually, aged proteins will be recognized and brought to their demise with the help of particular proteins and structures. These processes are examined and compared in prokaryotes and eukaryotes. It will be the end of the protein, the end of our journey, and the end of the book.

The synthesis of the polypetide is not the end of the story. The protein is not ready for work yet. Several modifications must take place. First, as the polypeptide is synthesized and exits from the ribosome, it must span the membranes of the endoplasmic reticulum by undergoing a specialized sequence in the N-terminus of the polypeptide, the signal peptide, and other proteins that act as recognition proteins or receptors. Eventually, the signal peptide will be cleaved, and it will not be part of the mature protein. The polypeptide then must fold to assume its correct three-dimensional structure, which, as this book has clearly indicated, is the ultimate quality control that will lead to the correct function of the protein. Correct folding, therefore, is paramount for the properties of the protein and constitutes a posttranslational regulation. Also, the folded protein can undergo other modifications, such as the addition of sugars or lipids, which in many cases are very important for function. As is true of everything else, the proteins do not live forever, and they eventually die by degradation mediated by specific molecular machines. Degradation of proteins is also an important component of regulation (the end point in the chain of regulation) because if a protein were allowed to live without control, cell or tissue gene regulation would be in jeopardy. Let us now examine these steps in protein maturation and degradation, emphasizing the 3-D level.

THE SIGNAL PEPTIDE AND ITS RECOGNITION

As mentioned earlier, the proteins are targeted either to the membranes of the endoplasmic reticulum (eukaryotes) or to the plasma membranes (prokaryotes). This is made possible by a portion in the N-terminus, the so-called signal peptide. Depending on the protein, the signal peptide can be 15 to 30 amino acids long and is characterized by a short segment of charged amino acids in the beginning followed by a hydrophobic core. Consider the following examples of some signal peptides:

Proalbumin: MKWVT*FLLLLFISGSAFS*R
IgG light chain: MDMRAPAQ*IFGFLLLLLFPGTRC*
Lysozyme: MRS*LLILVLCFLPLAALG*K

As we can see, the signal peptide does not have a consensus sequence with residues at similar positions. Also, the hydrophobic core (bold and italics) is located at various positions with respect to the starting methionine, and the length is not always the same. What is common among all signal peptides is the hydrophobic region, which is dominated by hydrophobic amino acids, mainly leucines, phenylalanines, and alanines. This makes sense when we consider that the signal peptide must span a membrane. Also, the signal peptide is cleaved while the polypeptide is still synthesized in the ribosome. This means that the process of translocation and targeting is intimately linked to protein synthesis and to the ribosome and that this process is cotranslational. As the signal-peptide-containing N-terminus of the proteins exits from the ribosome, the signal peptide is recognized by the signal recognition particle (SRP). The complex with SRP is then directed to the membrane via the interaction with the SRP receptor. The SRP-SRP receptor interaction is stabilized by GTP binding to the SRP-SRP receptor. As soon as GTP is bound by the complex, the nascent polypeptide dissociates from the SRP and is directed to the translocon, which

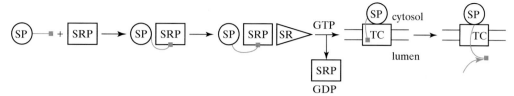

Figure 11.1. The sequence in protein translocation. SRP can bind the signal peptide (SP) as it emerges from the ribosome. The complex then binds with the SRP receptor (SR), which is associated with the translocon (TC). Upon binding of GTP, the SRP dissociates from the nascent polypeptide, which is then directed to the translocon. Upon GTP hydrolysis, SRP dissociates from the receptor.

spans the membrane, and protein synthesis continues (Figure 11.1). The final cleavage of the signal peptide is mediated by the enzyme signal peptidase as it exits from the translocon.

STRUCTURE AND FUNCTION OF THE SIGNAL RECOGNITION PARTICLE

SRP must recognize all signal peptides; therefore, because of the lack of homology, the recognition must be hydrophobic. The prokaryotic SRP consists of the 4.5S RNA and the Ffh protein. The Ffh protein contains three domains. The N-domain is a four-helix bundle and is associated with the G-domain, which is a Ras-like guanosine triphosphatase (GTPase). The G-domain is responsible for mediating the interaction between SRP and its receptor and regulating its function via the hydrolysis of GTP. The third domain is called the M-domain because it contains methionines and recognition sites for the signal peptide and the 4.5S RNA. The eukaryotic SRP is composed of a 300-nucleotide RNA (7S RNA) and six proteins and is divided into two domains. The S-domain contains four proteins – SRP54, SRP19, SRP72, and SRP68 – and the Alu domain contains two proteins – SRP9 and SRP 14. The two domains are linked by the SRP RNA, which has homology to the Alu repetitive sequences (Figure 11.2). The Alu domain is necessary for the elongation arrest of nascent chains after their signal sequences have been bound by the S-domain. The Alu RNA tertiary structure is crucial for transcription, maturation, nucleolus localization, and transport of SRP RNA. The SRP54 is the equivalent of the prokaryotic Ffh. All the components of the prokaryotic and eukaryotic SRP and SRP receptors share considerable homology, and they can, in fact, stay functional after exchange of their parts. For example, SRP54 can bind 4.5S RNA and hydrolyze GTP, and Ffh can replace SRP54, which results in a functional SRP.

The 3-D structure of the M-domain with the interacting portion of the 4.5S RNA has been solved and has provided interesting clues about the interaction with the signal peptide. The M-domain interacts with domain IV of the 4.5S RNA, which is nearly 50 nucleotides in length. This domain is also very similar

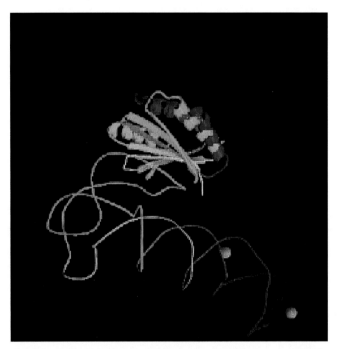

Figure 11.2. The 3-D structure of the human SRP Alu domain. SRP9 is composed of two helices (blue and dark green) and three beta strands (light blue) and interacts with SRP14 (rest of the structure). The bound SRP RNA is also shown with the characteristic U-turn. Weichenrieder et al., PDB file 1E8S, Nature 408: 167–73 (2000).

in the eukaryotic 7S. Domain IV is characterized by a tetraloop, a symmetric internal, and an asymmetric internal loop (Figure 11.3). Domain-M is a bundle of five helices, with helix 2 being noncanonical and, hence, subdivided to helix 2 and helix 2b. The M-domain recognizes the RNA at a minor groove by a HTH motif, which is composed of helices 2–4. The HTH motif presents two helices, 2, 2b and 3 to the minor groove, which make extensive direct and solvent-mediated contacts between conserved amino acids and nucleotides (Figure 11.3). We should note here that the HTH motif found in the M-domain is different from the one we encountered in transcriptional regulation where the HTH motif recognizes the major groove of DNA via one helix. The 4.5S RNA undergoes several conformational changes upon binding the M-domain. When the 4.5S RNA is free, the nucleotides of the asymmetric loops are stacked, but upon binding they they are extruded.

The RNA-protein interactions are facilitated by a network of waters and metal ions. We can see these interactions in Figure 11.4. In Figure 11.4A, the interactions of helices 2 and 3 with the symmetric loop are shown, and in Figures 11.4B and 11.4C, the interactions between helix 3 and the asymmetric loop are shown. Obviously, the interactions are dominated by numerous water molecules and potassium and magnesium ions.

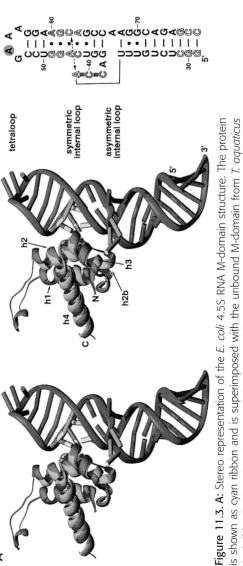

Figure 11.3. A: Stereo representation of the *E. coli* 4.5S RNA M-domain structure. The protein is shown as cyan ribbon and is superimposed with the unbound M-domain from *T. aquaticus* (red ribbon). Note that the proposed signal peptide recognition site (a loop connecting helices 1 and 2) was not determined in the *E. coli* structure (because of disorder probably due to the flexibility of this region). Region IV of the 4.5S RNA is blue with universally conserved and highly conserved nucleotides colored yellow and green, respectively. **B:** 3-D structure of unbound region IV, showing the differences in conformation when M-domain is not interacting. J. A. Dounda, *Science* 287: 1232–9 (2000). Reprinted with permission from American Association for the Advancement of Science.

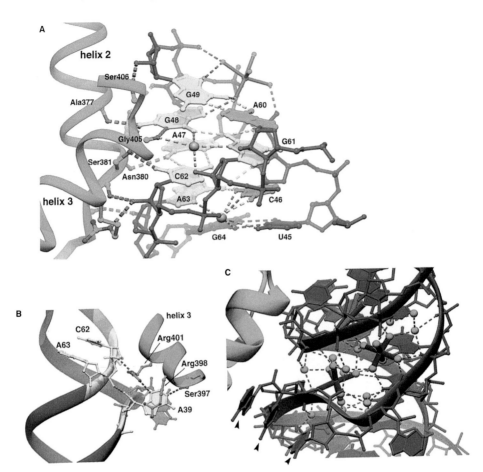

Figure 11.4. A: Protein-RNA interactions in the symmetric internal loop. The RNA backbone and helices 2 and 3 are blue. The 2′-OH groups are red spheres, orange spheres represent potassium, and red spheres with rods are hydrated magnesium ions. **B:** Protein-RNA interactions in the asymmetric internal loop. Note the stacking of A39, C40, and C41 and how Arg-398 virtually caps them. **C:** Twenty-eight water molecules (cyan spheres), two magnesium ions (red), and a potassium ion (orange) mediate hydrogen bonding (magenta dashes) in the asymmetric loop. The stacked A39, C40, and C41 are noted (arrowheads) for comparison with B. J. A. Dounda, Science 287: 1232–9 (2000). Reprinted with permission from American Association for the Advancement of Science.

The proposed signal-peptide-binding site is a groove, which is made up from the protein (the hydrophobic disordered loop between helix 1 and helix 2) and the RNA backbone adjacent to the symmetric internal loop (Figure 11.5A). In fact, the loop is similar to a signal peptide; it is composed of hydrophobic residues of similar length to signal peptide. The groove would be able to accommodate the signal peptide through hydrophobic interactions and electrostatic contacts. Actually, in some crystals, it has been found that the recognition groove can accommodate the loop from another M-domain (Figure 11.5B),

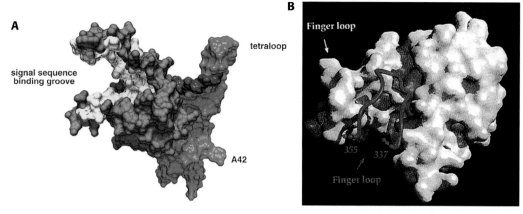

Figure 11.5. A: Surface representation of the 4.5S RNA M-domain structure oriented to visualize the proposed signal peptide recognition groove. The RNA is blue, the protein is magenta, the hydrophobic residues in the groove are yellow, and the phosphates in the RNA involved in the binding are red. J. A. Dounda, Science 287: 1232–9 (2000). **B:** The recognition groove can accommodate another loop finger in the *T. aquaticus* M-domain. R. J. Keenan, Cell 94: 181–91 (1998). A reprinted with permission from American Association for the Advancement of Science. B reprinted with permission from Elsevier Science.

which shows how the interaction with the signal peptide might occur. The hydrophobic portion of the signal peptide interacts with the hydrophobic region of the groove, while the positively charged amino acids of the signal peptide could interact with the phosphates in the RNA backbone.

THE BARREL OF BIRTH...

Protein folding is a rather complicated process that can take place in many different ways and involve many factors. Figure 11.6 depicts the processes of folding in prokaryotes and in eukaryotes. A quick glance reveals that there are many similarities; however, there are also some differences. As the polypeptide emerges from the ribosome, it is bound by the trigger factor (TF). The TF also binds the large subunit of translating and nontranslating ribosome. As the protein is further synthesized, it is folded either unassisted or assisted by chaperones. The chaperone-induced folding can be divided into two groups. In one group, proteins are bound by chaperones DnaK and DnaJ, which assist in cotranslational folding, and GrpE, which assists in posttranslational folding. In the other group, proteins are folded posttranlationally via the help of the barrel-like GroEL and GroES chaperones. All these folding reactions require ATP. In eukaryotes, folding is achieved by similar pathways, and the chaperones involved are similar to the prokaryotic counterparts in sequence and structure. In yeast and mammalian cells, instead of TF, there is the nascent chain-associated complex (NAC) that cross-links to the emerging polypeptide, as short as 17 residues. Such binding might protect the nascent chains from degradation. In yeast, Ssb

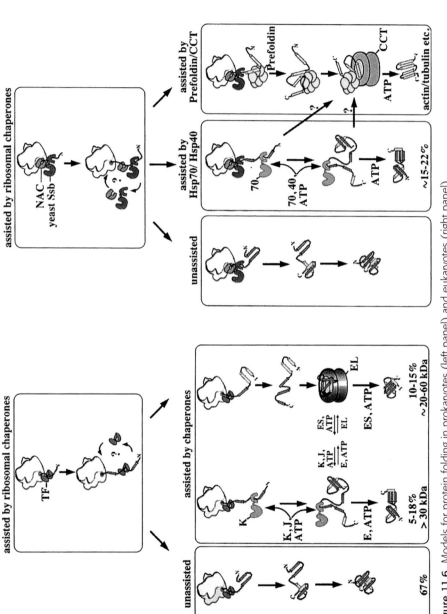

Figure 11.6. Models for protein folding in prokaryotes (left panel) and eukaryotes (right panel). B. Bukau and E. Craig, Cell 101:119–22 (2000). Reprinted with permission from Elsevier Science.

proteins (which are homologs of Hsp70) have also been found to cross-link with nascent chains and with ribosomes. The chaperone-assisted folding can be achieved either by the Hsp70/Hsp40 chaperones or by the prefoldin/CCT chaperones. Hsp is the abbreviation for heat shock protein because they were identified first after heat shock of the cells. Hsp70 is structurally related to DnaK and Hsp40 to DnaJ. CCT is the barrel-like structure similar to GroEL. Likewise, folding of the eukaryotic proteins requires ATP.

In folding a protein, the general idea is to bury the hydrophobic residues inside the protein and to expose the hydrophilic ones to the outside. Therefore, all the proteins that are assisting in folding must operate with this rule. Next we will examine the 3-D structure of the major players in folding, which will provide very important insights for the process.

The 3-D Structure of DnaK and Hsp70

DnaK acts as follows. First, ATP binds DnaK. At this state, DnaK can bind peptides weakly and release them. The cochaperone DnaJ stimulates hydrolysis of ATP by DnaK, resulting in DnaK-ADP, which binds peptides tightly. For the release of the peptides, dissociation of ADP is necessary, and ADP is catalyzed by GrpE with ATP rebinding to DnaK. We will now examine the structure and association between DnaK, DnaJ, and GrpE.

DnaK and its eukaryotic homolog Hsp70 consist of three domains. The N-terminal domain has been shown to be the one with the ATPase activity. This domain is followed by the substrate domain, which binds the polypeptides. The third domain is involved with interactions with the cochaperone DnaJ (for DnaK) or Hsp40 (for Hsp70). The substrate-binding site of *E. coli* DnaK and of rat Hsp70 have been solved by crystallography and NMR, respectively. The substrate-binding sites are remarkably similar in both the prokaryotic and the eukaryotic proteins, and they are made of an eight-beta-strand sheet (Figure 11.7). However, a major difference is that DnaK has a variable domain, which is not necessary for coupling the ATPase and substrate domains. The position of the helix is also different. The helix of the DnaK sits on the top surfaces of loops L4,5 and L1,2, while the helix in Hsp70 runs in a cleft along these loops. Also, the Hsp70 helix is kinked and considered as two helices. The substrate binding is in a channel formed by L1,2 and L3,4. The substrate for DnaK is shown in Figure 11.7A as a red line and for Hsp70 is the C-terminus peptide of the domain (Figure 11.7B). The peptides uses a specific leucine, which resides in a well-formed hydrophobic pocket. Leu-539 is bound by intramolecular interactions in a hydrophobic groove, especially in the structure of Hsp70 where the C-terminus of the domain is in the substrate-binding site. DnaK and Hsp70 bind short peptides composed of clusters of hyrdophobic amino acids, which are flanked by basic residues. The ATPase domain of DnaK will be presented with the structure of GrpE (see later).

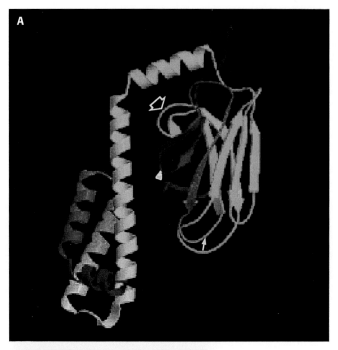

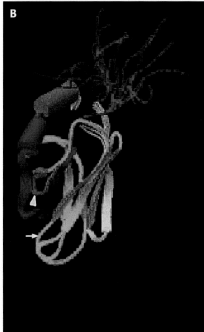

Figure 11.7. Comparison of the substrate-binding site from the *E. coli* DnaK (A) and from the rat Hsp70 (B). Note the structural similarity in the beta-sheet region. Note also the variable region in DnaK that is absent in Hsp70 and the difference in the orientation of the helix. The substrate is depicted as a red line in A. L1,2 is denoted by an arrowhead, L3,4 by an arrow, and L4,5 is marked with an open arrowhead. **A:** Zhu et al., PDB file 1DKX, Science 272: 1606–14 (1996). **B:** Morshauser et al., PDB file 1CKR, J. Mol. Biol. 289: 1387–403 (1999).

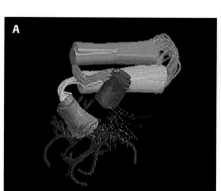
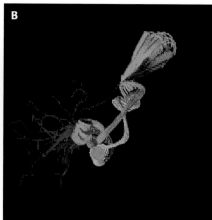

Figure 11.8. A: The J-domain is made up of four alpha helices. K. Huang et al., PDB file 1BQ0, Protein Sci. 8: 203–14 (1999). **B:** The V-shaped CR domain. The zinc ions are shown as gray spheres. M. Martinez-Yamout et al., PDB file 1EXK, J. Mol. Biol. 300: 805–18 (2000).

The 3-D Structure of DnaJ

Members of the DnaJ family contain one to four domains and can be divided in different types depending on this number. The common domain in all of them is the N-terminus domain, otherwise called the J-domain. The J-domain is followed by the G/F-domain, which is rich in glycines and phenylalanines. Then the cysteine-rich domain (CR) follows. The C-terminal domain is the least conserved of all. The J- and G/F-domains contain residues necessary for interaction with DnaK. The CR domain is thought to be involved in substrate binding, but it is not by itself sufficient for folding. Obviously, other domains might be involved. The 3-D structure of the J-domain reveals a four-alpha-helix bundle. The CR domain is shaped as an extended V and has two zinc-binding sites (Figure 11.8). In eukaryotes, the homologous protein to DnaJ is the Hsp40, which interacts and functions together with the Hsp70, the DnaK homolog.

The 3-D Structure of GrpE

GrpE binds the ATPase domain of DnaK as a dimer, even though all the interactions involve only one monomer. GrpE monomer is made up of a long alpha helix, which forms the dimer interface. The helix leads to a small two-helix bundle. A beta-sheet domain made up of six beta strands emanates from the C-terminal end of the two-helix bundle. The beta domain has also a limited hydrophobic core. The ATPase domain of DnaK consists of two large domains, A and B, which are divided into subdomains I and II. The nucleotide-binding cleft is between subdomains IIA and IA (Figure 11.9). The ATPase domains between DnaK and its eukaryotic counterpart, Hsp70, are quite similar. The major areas of interaction between GrpE and the ATPase domain are

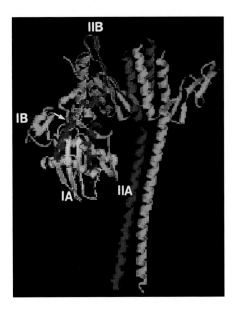

Figure 11.9. The 3-D structure of *E.coli* GrpE complex with the ATPase domain of *E.coli* DnaK. GrpE is a dimer (right structure; blue and green), and DnaK is a monomer (left structure). An arrow indicates the nucleotide-binding cleft between subdomains IIA and IA. Harrison et al., PDB file 1DKG, Science 276: 431–5 (1997).

between the two faces of the beta-sheet domain of one GrpE monomer (called the proximal monomer) and domains IB and IIB, which are on both sides of the nucleotide-binding cleft. Other contacts involve the long helix. The elongated structure of GrpE suggests possible interactions with the peptide-binding domain of DnaK. It seems that the positioning of the ATPase domain would also place the C-terminal peptide-binding domain of DnaK near the N-terminus of the long GrpE helices. It is possible that the N-terminus of the long helices is involved in peptide release.

The 3-D Structure of Prefoldin

As mentioned earlier, prefoldin interacts with the nascent polypeptide chains and stabilizes nonnative proteins for subsequent folding in the cavity of a chaperone. Prefoldin is present in eukaryotes and archaea and is composed of two related classes of subunits that form a hexamer. The best way to describe prefoldin is as a jellyfish-like structure, where six tentacles (two alpha and four beta subunits) are protruding from a double beta barrel. Each tentacle (alpha or beta subunit) is a coiled coil (Figure 11.10). The distal regions of the coiled coil consist of hydrophobic residues that are required for binding nonnative proteins. The requirement of these regions has been documented in experiments where these regions were deleted, resulting in an inability to bind polypeptides.

The 3-D Structure of GroEL and GroEs

GroEL and its eukaryotic counterpart Hsp60 are shaped like a barrel. The 3-D structure of GroEL has been solved with or without GroES. First, a

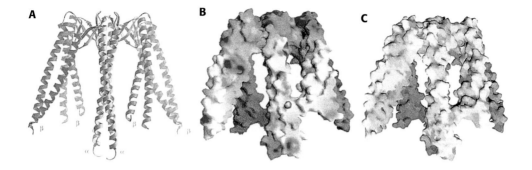

Figure 11.10. The 3-D structure of prefoldin from *Methanobacterium thermoautotrophicum* (archaeon). **A:** A ribbon model indicating the six coiled coils and the beta barrel on the top. The alpha subunits are orange, and the beta subunits are light blue. **B:** Same as in A but the surface of prefoldin is shown with the electrostatic potential (red acidic, blue basic residues). **C:** Same as in B but the hydrophobic patches (yellow) are shown at the distal tips, which are implicated in binding to the nonnative proteins. I. Moarefi, Cell 103: 621–32 (2000). Reprinted with permission from Elsevier Science.

protein is brought into GroEL. The protein is brought by DnaK, which then dissociates. ATP hydrolysis releases the protein into the cavity. GroEL binds to short amphipathic peptides that adopt an alpha helical conformation. The substrate binding is mediated by hydrophobic interactions. When the protein is inside GroEL, GroES caps GroEL, which initiates folding. The scenario is that ATP hydrolysis along with binding of GroES allows certain distortion in the inside of GroEL so that the hydrophilic residues are exposed in the inside of the barrel. The unfolded protein buries its hydrophobic residues and also exposes the hydrophilic. This helps the protein to fold correctly. The barrel consists of four rings, each ring consisting of seven identical subunits. The building subunit is made up of three domains – the apical, the intermediate and the equatorial. Upon binding of ATP (on the equatorial domain) the intermediate domain folds downward, toward the equatorial domain. The apical domain twists clockwise, and the equatorial domain twists counterclockwise. Then, the cochaperone GroES binds, resulting in motion of the apical domain, which rotates nearly 90 degrees. These alterations in the basic subunit of GroEL result in exposing hydrophilic residues in the interior of the barrel. The unfolded protein buries its hydrophobic residues in its interior and exposes its hydrophilic residues, resulting in folding. The protein is then released. However, if it is not properly folded, it will be degraded or will undergo another cycle of folding (Figure 11.11).

Another way to fold a protein is by the disulfide-bonding method, where bonds are formed between the sulfur of two cysteines. All proteins contain cysteines, but not all cysteines form disulfide bonds. The enzyme that is responsible for correct disulfide bonding is called protein disulfide isomerase (PDI). This

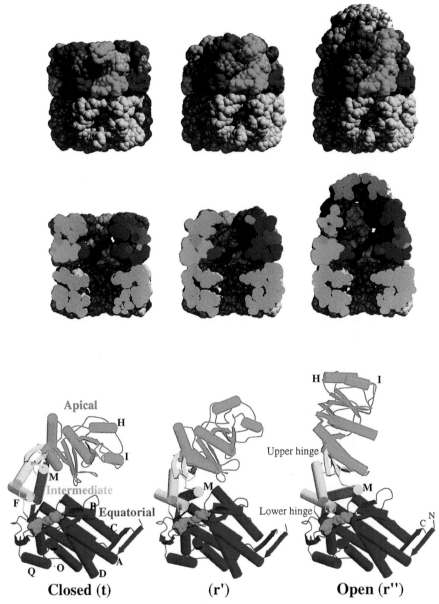

Figure 11.11. Top and middle rows: spacefill models of GroEL showing the two-ring arrangement of the 14 subunits (left). The same model is in the middle, but it has the observed distortions due to ATP binding. The structure after GroES (gray) binding appears on the right. In the lower row, we can see the alteration in the 3-D structure of the building subunit (see text). M. Karplus, J. Mol. Biol. 302: 303–13 (2000). Reprinted with permission from Academic Press Ltd.

enzyme contains cysteines that can interact randomly with any cysteine of the protein to be folded. If the bonding of the enzyme promotes rearrangement of the disulfide bonding in the protein, which in turn results in a more stable pattern, PDI dissociates. After several steps, the correct disulfide bonds will result in a correctly folded protein. PDI might also foster the aggregation of misfolded proteins when the amount of them exceeds the chaperone capacity. This clustering might save the protein from degradation and another trial at folding.

... AND THE BARREL OF DEATH

Proteins cannot live forever. This would be against any regulation. If proteins were allowed to live forever, regulation of transcription would not be necessary. This lack of regulation could be a problem when we want specificity in space and time. Eventually, proteins must die by proteolysis. Proteolysis is also important when proteins are misfolded. A misfolded protein must be degraded or go through another cycle of folding. ATP-dependent proteases are shaped like barrels and are quite similar to GroEL chaperone. In Figure 11.12, we can compare the proteolysis pathways in prokaryotes and eukaryotes. In prokaryotes, the protease is made up of one protease compartment and two ATPase compartments. The protease compartment of *E. coli* ClpAP is made up of two rings with seven subunits per ring (GroEL has four rings). The substrate binds the ATPase compartment and is internalized in the protease

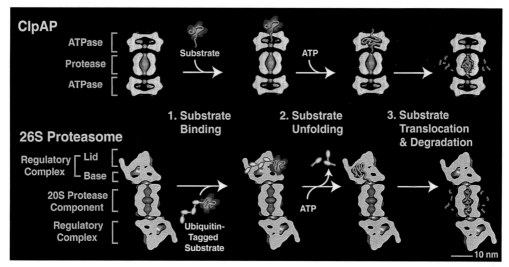

Figure 11.12. ATP-dependent proteolysis in prokaryotes (top panel; by *E. coli* ClpAP) and eukaryotes (lower panel; 26S proteasome). See text for details. M. E. Gottesman, Science 286: 1888–93 (1999). Reprinted with permission from American Association for the Advancement of Science.

A

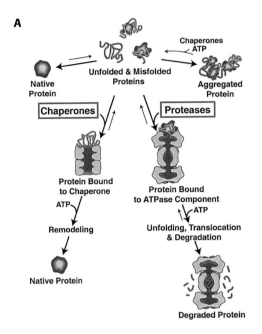

B

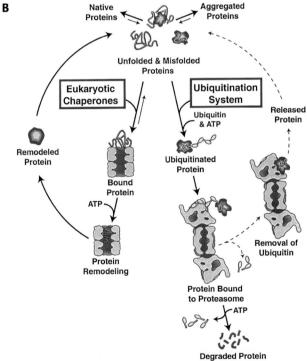

Figure 11.13. Quality control of proteins in prokaryotes (A) and in eukaryotes (B). Interplay between chaperones and proteases in folding and degradation of proteins. M. E. Gottesman, Science 286: 1888–93 (1999). Reprinted with permission from American Association for the Advancement of Science.

compartment where it is degraded. Hydrophobic clusters flanked by ba-sic residues are the targets for substrate binding. Another *E. coli* protease, HslVU has two rings and six subunits in each ring. In eukaryotes, the ATP-dependent proteases that depend also on ubiquitin are called proteasomes. Many ubiquitin molecules form polyubiquitin chains that recognize and mod-ify protein. For this, ubiquitin-conjugated enzymes and ubiquitin ligases are responsible. Proteasomes have three compartments: two regulatory and the 20S protease (core) component, which is shaped like a barrel. The top regula-tory compartment is made up of the lid, which contains proteins that recognize ubiquitin, and the base, which contains proteins with possible signal transduc-tion function and ubiquitin isopeptidase. The 20S core is made up of four rings with seven subunits per ring. The top and the bottom ring are made up of alpha subunits, and the two central rings are composed of beta subunits. This arrangement divides the lumen into three chambers. The peptidase active sites are in the central chamber, created between the two central rings (Figure 11.12). Ubiquitinizing is necessary for degradation by the proteasome. Ubiquity binds the substrate and brings it to the regulatory complex. Then, ubiquitin dissociates, and the substrate is internalized and degraded. The eukaryotic pro-teasomes are threonine proteases and belong to a new family named Ntn for N-terminal nucleophile. The N-terminal threonine is the nucleophile that me-diates catalysis. The proteasome from the archaeon *Thermoplasma* is also a threonine protease. The same is true for the *E. coli* HslVU, but not for ClpAP, which is a serine protease. Other proteases are ATP independent. These could look like barrels, or they might be structured differently (such as tricons).

In addition to their role in the death of proteins, proteases and proteasome also degrade proteins that have not been able to fold normally. This quality control can be seen in Figure 11.13 for both prokaryotes and eukaryotes. Note here the great role that ubiquitinization has on gene regulation. Ubiquitin rec-ognizes proteins that have specific roles in regulation and by ubiquitinizing them leads them to their degradation. In this sense, proteolysis regulates a variety of cell functions. Some examples are cytokine-induced gene expression (specific ubiquitinization of IkB and activation of NF-kB) or cell cycle regulation.

Further Reading and References

BOOKS

Branden, C., and Tooze, J. (1999). Introduction to protein structure, Garland, New York.

Lewin, B. (2000). Genes VII, Oxford, New York.

Lodish, H., Berk, A., Zipursky, S. L., Matsudaira, P., Baltimore, D., and Darnell, J. (2000). Molecular cellular biology, W. H. Freeman, New York.

Ptashne, M. (1992). A genetic switch, Cell Press and Blackwell Scientific Publications, Cambridge, MA.

Singer, M., and Berg, P. (1991). Genes and genomes. University Science Books, Mill Valley, CA.

Weaver, R. F. (1999). Molecular biology, McGraw-Hill, New York.

SCIENTIFIC PAPERS AND REVIEWS

Anderson, J. E., Ptashne, M., and Harrison, S. C. (1987). Structure of the repressor-operator complex of bacteriophage 434. Nature 326: 846–52.

Andrews, B. J., and Donoviel, M. S. (1995). A heterodimeric transcriptional repressor becomes crystal clear. Science 270: 251–3.

Antson, A. A., Dodson, E. J., Dodson, G., Greaves, R. B., Chen, X., and Gollnick, P. (1999). Structure of the trp RNA-binding attenuation protein, TRAP, bound to RNA. Nature 401: 235–42.

Arents, G., and Moudrianakis, E. N. (1995). The histone fold: A ubiquitous architectural motif utilized in DNA compaction and protein dimerization. Proc. Natl. Acad. Sci. USA 92: 11170–4.

Asturias, F. J., Jiang, Y. W., Myers, L. C., Gustafsson, C. M., and Kornberg, R. D. (1999). Conserved structures of mediator and RNA polymerase II holoenzyme. Science 283: 985–7.

Ban, N., Freeborn, B., Nissen, P., Penczek, P., Grassucci, R. A., Sweet, R., Frank, J., Moore P. B., and Steitz, T. A. (1998). A 9 Å resolution X-Ray crystallographic map of the large ribosomal subunit. Cell 93: 1105–15.

Ban, N., Nissen, P., Hansen, J., Capel, M., Moore, P. B., and Steitz, T. A. (1999). Placement of protein and RNA structures into a 5 A-resolution map of the 50S ribosomal subunit. Nature 400: 841–7.

Ban, N., Nissen, P., Hansen, J., Moore, P. B., and Steitz, T. A. (2000). The complete atomic structure of the large ribosomal subunit at 2.4 Å resolution. Science 289: 905–19.

Bass, B. L. (2000). Double-stranded RNA as a template for gene silencing. Cell 101: 235–8.

Batey, R. T., Rambo, R. P., Lucast, L., Rha, B., and Doudna, J. A. (2000). Crystal structure of the ribonucleoprotein core of the signal recognition particle. Science 287: 1232–9.

Battiste, J. L., Pestova, T. V., Hellen, C. U., and Wagner, G. (2000). The eIF1A solution structure reveals a large RNA-binding surface important for scanning function. Mol. Cell 5: 109–19.

Baumeister, W., Walz, J., Zuhl, F., and Seemuller, E. (1998). The proteasome: Paradigm of a self-compartmentalizing protease. Cell 92: 367–80.

Beamer, L. J., and Pabo, C. O. (1992). Refined 1.8 Angstrom crystal structure of the lambda repressor-operator complex. J. Mol. Biol. 227: 177–96.

Becker, S., Groner, B., and Muller, C. W. (1998). Three-dimensional structure of the Stat3β homodimer bound to DNA. Nature 394: 145–51.

Bell, C. E., Frescura, P., Hochschild, A., and Lewis, M. (2000). Crystal structure of the λ repressor C-terminal domain provides a model for cooperative operator binding. Cell 101: 801–11.

Berger, J. M., Gamblin, S. J., Harrison, S. C., and Wang, J. C. (1996). Structure and mechanism of DNA topoisomerase II. Nature 379: 225–32.

Biou, V., Shu, F., and Ramakrishnan, V. (1995). X-ray crystallography shows that translational initation factor IF3 consists of two compact α/β domains linked by an α-helix. EMBO J. 14: 4056–64.

Birck, C., Poch, O., Romier, C., Ruff, M., Mengus, G., Lavigne, A., Davidson, I., and Moras, D. (1998). Human TAFII28 and TAFII18 interact through a histone fold encoded by atypical evolutionary conserved motifs also found in the SPT3 family. Cell 94: 239–49.

Blum, B., Bakalara, N., and Simpson, L. (1990). A model for RNA editing in kinetoplasmid mitochondria: "Guide" RNA molecules transcribed from maxicircle DNA provide the edited information. Cell 60: 189–98.

Bochkarev, A., Pfuetzner, R. A., Edwards, A. M., and Frappier, L. (1997). Structure of the single-stranded-DNA-binding domain of replication protein A bound to DNA. Nature 385: 176–81.

Bogden, C. E., Fass, D., Bergman, N., Nichols, M. D., and Berger, J. M. (1999). The structural basis for terminator recognition by the Rho transcription termination factor. Mol. Cell 3: 487–93.

Brino, L., Urzhumtsev, A., Mousli, M., Bronner, C., Mitschler, A., Oudet, P., and Moras, D. (2000). Dimerization of Escherichia coli DNA-gyrase B provides a structural mechanism for activating the ATPase catalytic center. J. Biol. Chem. 275(13): 9468–75.

Brodersen, D. E., Clemons, W. M., Jr., Carter, A. P., Morgan-Warren, R. J., Wimberly, B. T., and Ramakrishnan, V. (2000). The structural basis for the action of the antibiotics tetracycline, pactamycin, and hygromycin B on the 30S ribosomal subunit. Cell 103: 1143–54.

Bukau, B., and Horwich, A. L. (1998). The Hsp70 and Hsp60 chaperone machines. Cell 92: 351–66.

Bukau, B., Deuerling, E., Pfund, C., and Craig, E. A. (2000). Getting newly synthesized proteins into shape. Cell 101: 119–22.

Caprara, M. G., Lehnert, V., Lambowitz, A. M., and Westhof, E. (1996). A tyrosyl-tRNA synthetase recognizes a conserved tRNA-like structural motif in the group I intron catalytic core. Cell 87: 1135–45.

Carrodeguas, J. A., Theis, K., Bogenhagen, D. F., and Kisker, C. (2001). Crystal structure and deletion analysis show that the accessory subunit of mammalian DNA polymerase gamma, PolgammaB, functions as a homodimer. Mol. Cell 7: 43–54.

Carter, A. P., Clemons, W. M., Brodersen, D. E., Morgan-Warren, R. J., Hartsch, T., Wimberly, B. T., and Ramakrishnan, V. (2001). Atomic structure of an initiation factor bound to the 30S ribosomal subunit. Science 291: 498–501.

Carter, A. P., Clemmons, W. M., Brodersen, D. E., Morgan-Warren, R. J., Wimberly, B. T., and Ramakrishnan, V. (2000). Functional insights from the structure of the 30S ribosomal subunit and its interactions with antibiotics. Nature 407: 340–8.

Cate, J. H., Yusupov, M. M., Yusupova, G. Z., Earnest, T. N., and Noller, H. F. (1999). X-ray crystal structures of 70S ribosome functional complexes. Science 285: 2095–104.

Celander, D. W., and Cech, T. R. (1991). Visualizing the higher order folding of a catalytic RNA molecule. Science 251: 401–7.

Cheetham, G. M. T., and Steitz, T. A. (1999). Structure of a transcribing T7 RNA polymerase initiation complex. Science 286: 2305–9.

Chen, L., Glover, J. N. M., Hogan, P. G., Rao, A., and Harrison, S. C. (1998). Structure of the DNA-binding domains from NFAT, Fos and Jun bound specifically to DNA. Nature 392: 42–8.

Cho, H., Ha, N., Kang, L., Chung, K., Back, S., Jang, S., and Oh, B. (1998). Crystal structure of RNA helicase from genotype 1b hepatitis C virus. J. Biol. Chem. 273(24): 15045–52.

Copertino, D. W., and Hallick, R. B. (1993). Group II and group III introns of twintrons: Potential relationships with nuclear pre-mRNA introns. Trends Biochem. Sci. 18: 467–71.

Cramer, P., Bushnell, D. A., Fu, J., Gnatt, A. L., Maier-Davis, B., Thompson, N. E., Burgess, R. R., Edwards, A. M., David, P. R., and Kornberg, R. D. (2000). Architecture of RNA polymerase II and implications for the transcription mechanism. Science 288: 640–9.

Darst, S. A., Edwards, A. M., Kubalek, E. W., and Kornberg, R. D. (1991). Three-dimensional structure of yeast RNA polymerase II at 16 A resolution. Cell 66: 121–8.

Davenport, R. J., Wuite, G. J. L., Landick, R., and Bustamante, C. (2000). Single-molecule study of transcriptional pausing and arrest by E. coli RNA polymerase. Science 287: 2497–500.

Decanniere, K., Babu, A. M., Reeve, J. N., and Heinemann, U. (2000). Crystal structures of recombinant Hmfa and Hmfb from the hyperthermophilic archaeon methanothermus ferridus. J. Mol. Biol. 303: 35–47.

Dernburg, A. F., Broman, K. W., Fung, J. C., Marshall, W. F., Philips, J., Agard, D. A., and Sedat, J. W. (1996). Perturbation of nuclear architecture by long-distance chromosome interactions. Cell 85: 745–59.

Deo, R. C., Bonanno, J. B., Sonenberg, N., and Burley, S. K. (1999). Recognition of polyadenylate RNA by the poly(A)-binding protein. Cell 98: 835–45.

Doublie, S., Tabor, S., Long, A. M., Richardson, C. C., and Ellenberger, T. (1998). Crystal structure of a bacteriophage T7 DNA replication complex at 2.2 Å resolution. Nature 391: 251–7.

Ellenberger, T. E., Brandl, C. J., Struhl, K., and Harrison, S. C. (1992). The GCN4 basic region leucine zipper binds DNA as a dimer of uninterrupted alpha helices: Crystal structure of the protein-DNA complex. Cell 71: 1223–37.

Elrod-Erickson, M., Benson, T. E., and Pabo, C. O. (1998). High-resolution structures of vatiant Zif268-DNA complexes: Implications for understanding zinc finger-DNA recognition. Structure 6: 451–64.

Erwin, D., Valentine, J., and Jablonski, D. (1997). The origin of animal body plans. Amer. Sci. 85: 126–37.

Escalante, C. R., Yie, J., Thanos, D., and Aggarwal, A. K. (1998). Structure of IRF-1 with bound DNA reveals determinants of interferon regulation. Nature 391: 103–6.

Fabrera, C., Farrow, M. A., Mukhopadhyay, B., de Crecy-Lagard, V., Ortiz, A. R., and Schimmel, P. (2001). An aminoacyl tRNA synthetase whose sequence fits into neither of the two known classes. Nature 411: 110–4.

Feagin, J. E., Abraham, J. M., and Stuart, K. (1988). Extensive editing of the cytochrome c oxidase III transcript in Trypanosoma brucei. Cell 53: 413–22.

Femino, A. M., Fay, F. S., Fogarty, K., and Singer, R. H. (1998). Visualization of single RNA transcripts in situ. Science 280: 585–90.

Festenstein, R., Tolaini, M., Corbella, P., Mamalaki, C., Parrington, J., Fox, M., Miliou, A., Jones, M., and Kioussis, D. (1996). Locus control region function and heterochromatin-induced position effect variegation. Science 271: 1123–5.

Filipski, J., Leblanc, J., Youdale, T., Sikorska, M., and Walker, P. R. (1990). Periodicity of DNA folding in higher order chromatin structures. EMBO J. 9(4): 1319–27.

Fire, A. (1999). RNA-triggered gene silencing. Trends Genet. 15: 358–63.

Fletcher, C. M., Pestova, T. V., Hellen, C. U., and Wagner, G. (1999). Structure and interactions of the translation initiation factor eIF1. EMBO J. 18: 2631–7.

Frank, J. (1998). How the ribosome works. Amer. Sci. 86: 428–39.

Franklin, M. C., Wang, J., and Steitz, T. A. (2001). Structure of the replicating complex of a Pol α family DNA polymerase. Cell 105: 657–67.

Gabashvili, I. S., Agrawal, R. K., Spahn, C. M. T., Grassucci, R. A., Svergun, D. I., Frank, J., and Penczek, P. (2000). Solution structure of the E. coli 70S ribosome at 11.5 Å resolution. Cell 100: 537–49.

Glasfeld, A., Koehler, A. N., Schumacher, M. A., and Brennan, R. G. (1999). The role of lysine 55 in determining the specificity of the purine repressor for its operators through minor groove interactions. J. Mol. Biol. 291(2): 347–61.

Golden, B. L., Gooding, A. R., Podell, E. R., and Cech, T. R. (1998). A preorganized active site in the crystal structure of Tetrahymena ribozyme. Science 282: 259–64.

Greider, C. W. (1999). Telomeres do D-loop-T-loop. Cell 97: 419–22.

Griffith, J. D., Comeau, L., Rosenfield, S., Stansel, R. M., Bianchi, A., Moss, H., and de Lange, T. (1999). Mammalian telomeres end in a large duplex loop. Cell 97: 503–14.

Gulbis, J. M., Kelman, Z., Hurwitz, J., O'Donnell, M., and Kuriyan, J. (1996). Structure of the C-terminal region of p21 complexed with human PCNA. Cell 87: 297–306.

Hakansson, K., Doherty, A. J., Shuman, S., and Wigley, D. B. (1997). X-Ray crystallography reveals a large conformational change during guanyl transfer by mRNA capping enzyme. Cell 89: 545–53.

Hard, T., Kellenbach, E., Boelens, R., Maler, B. A., Dahlman, K., Freedman, L. P., Carlstedt-Duke, J., Yamamoto, K. R., Gustafsson, J., and Kaptein, R. (1990). Solution structure of the glucocorticoid receptor DNA-binding domain. Science 249: 157–60.

Harrison, C. J., Hayer-Hartl, M., Di Liberto, M., Hartl, F. U., and Kuriyan, J. (1997). Crystal structure of the nucleotide exchange factor GrpE bound to the ATPase domain of the molecular chaperone DnaK. Science 276: 431–5.

Hodel, A. E., Gershon, P. D., and Quiocho, F. A. (1998). Structural basis for sequencing-nonspecific recognition of 5′-capped mRNA by a cap-modifying enzyme. Mol. Cell 1: 443–7.

Hopkin, K. (1997). Spools, switches, or scaffolds: How might histones regulate transcription? J. NIH Res. 9: 34–7.

Horvath, M. P., Schweiker, V. L., Bevilacqua, J. M., Ruggles, J. A., and Schultz, S. C. (1998). Crystal structure of the Oxytricha nova telomere end binding protein complexed with single strand DNA. Cell 95: 963–74.

Hosfield, D. J., Mol, C. D., Shen, B., and Tainer, J. A. (1998). Structure of the DNA repair and replication endonuclease and exonuclease FEN-1: Coupling DNA and PCNA binding to FEN-1 activity. Cell 95: 135–46.

Howard, M. J. (1998). Protein NMR spectroscopy. Curr. Biol. 8(10): R331–3.

Ito, K., Uno, M., and Nakamura, Y. (2000). A tripeptide 'anticodon' deciphers stop codons in messenger RNA. Nature 403: 680–4.

Jin, Y., Mead, J., Li, T., Wolberger, C., and Vershon, A. K. (1995). Altered DNA recognition and bending by insertions in the α2 tail of the yeast a1/α2 homeodomain heterodimer. Science 270: 290–3.

Joseph, S., Weiser, B., and Noller, H. F. (1997). Mapping the inside of the ribosome with an RNA helical ruler. Science 278: 1093–8.

Kastner, B. (1998). Purification and electron microscopy of spliceosomal snRNPs. In RNP particles, splicing and autoimmune diseases (J. Scenkel, Ed.), Springer, Berlin, pp. 95–140.

Kambach, C., Walke, S., Young, R., Avis, J. M., de la Fortelle, E., Raker, V. A., Luhrmann, R., Li, J., and Nagai, K. (1999). Crystal structures of two Sm protein complexes and their implications for the assembly of the spliceosomal snRNPs. Cell 96: 375–87.

Kambach, C., Walke, S., and Nagai, K. (1999). Structure and assembly of the spliceosomal small nuclear ribonucleoprotein particles. Curr. Opin. Struct. Biol. 9: 222–30.

Kang, C., Zhang, X., Ratliff, R., Moyzis, R., and Rich, A. (1992). Crystal structure of four-stranded Oxytricha telomeric DNA. Nature 356: 126–31.

Keck, J. L., Roche, D. D., Lynch, A. S., and Berger, J. M. (2000). Structure of the RNA polymerase domain of E. coli primase. Science 287: 2482–6.

Keenan, R. J., Freymann, D. M., Walter, P., and Stroud, R. M. (1998). Crystal structure of the signal sequence binding subunit of the signal recognition particle. Cell 94: 181–91.

Kiefer, J. R., Mao, C., Braman, J. C., and Beese, L. S. (1998). Visualizing DNA replication in a catalytically active Bacillus DNA polymerase crystal. Nature 391: 304–7.

Kim, C. A., and Berg, J. M. (1996). A 2.2 Angstrom resolution crystal structure of a designed zinc finger protein bound to DNA. Nature Struct. Biol. 3: 940–5.

Kissinger, C. R., Liu, B., Martin-Blanco, E., Kornberg, T. B., and Pabo, C. O. (1990). Crystal structure of an engrailed homeodomain-DNA complex at 2.8 Å resolution: A framework for understanding homeodomain-DNA interactions. Cell 63: 579–90.

Kjeldgaard, M., Nissen, P., Thirup, S., and Nyborg, J. (1993). The crystal structure of elongation factor EF-Tu from Thermus aquaticus in the GTP conformation. Structure 1(1): 35–50.

Konforti, B. B., Abramovitz, D. L., Duarte, C. M., Karpeisky, A., Beigelan, L., and Pyle, A. M. (1998). Ribozyme catalysis from the major groove of group II intron domain 5. Mol. Cell 1: 433–41.

Konig, P., Giraldo, R., Chapman, L., and Rhodes, D. (1996). The crystal structure of the DNA-binding domain of yeast RAP1 in complex with telomeric DNA. Cell 85: 125–36.

Korolev, S., Hsieh, J., Gauss, G. H., Lohman, T. M., and Waksman, G. (1997). Major domain swiveling revealed by the crystal structures of complexes of E. coli rep helicase bound to single-stranded DNA and ADP. Cell 90: 635–47.

Korzheva, N., Mustaev, A., Kozlov, M., Malhotra, A., Nikiforov, V., Goldfarb, A., and Darst, S. A. (2000). A structural model of transcriptional elongation. Science 289: 619–25.

Larsen, C. N., and Finley, D. (1997). Protein translocation channels in the proteasome and other proteases. Cell 91: 431–4.

Lavoie, B. D., Shaw, G. S., Millner, A., and Chaconas, G. (1996). Anatomy of a flexer-DNA complex inside a higher-order transposition intermediate. Cell 85: 761–71.

Lawrence, J. B., Singer, R. H., and Marselle, L. M. (1989). Highly localized tracks of specific transcripts within interphase nuclei visualized by in situ hybridization. Cell 57: 493–502.

Leuther, K. K., Bushnell, D. A., and Kornberg, R. D. (1996). Two-dimensional crystal-lography of TFIIB- and IIE-RNA polymerase II complexes: Implications for start site selection and initiation complex formation. Cell 85: 773–9.

Levin, D. S., Bai, W., Yao, N., O'Donnel, M., and Tomkinson, A. E. (1997). An interaction between DNA ligase I and proliferating cell nuclear antigen: Implications for Okazaki fragment synthesis and joining. Proc. Natl. Acad. Sci. USA 94: 12863–8.

Lewis, M., Chang, G., Horton, N. C., Kercher, M. A., Pace, H. C., Schumacher, M. A., Brennan, R. G., and Lu, P. (1996). Crystal structure of the lactose operon repressor and its complexes with DNA and inducer. Science 271: 1247–54.

Li, T., Stark, M. R., Johnson, A. D., and Wolberger, C. (1995). Crystal structure of the MATa1/MATα2 homeodomain heterodimer bound to DNA. Science 270: 262–9.

Liao, S., Lin, J., Do, H., and Johnson, A. E. (1997). Both lumenal and cytosolic gating of the aqueous ER translocon pore are regulated from inside the ribosome during membrane protein integration. Cell 90: 31–41.

Lima, C. D., Wang, L. K., and Shuman, S. (1999). Structure and mechanism of yeast RNA triphosphatase: An essential component of the mRNA capping apparatus. Cell 99: 533–43.

Liu, D., Ishima, R., Tong, K. I., Bagby, S., Kokubo, T., Muhandiram, D. R., Kay, L. E., Nakatani, Y., and Ikura, M. (1998). Solution structure of a TBP-TAFII230 complex: Protein mimicry of the minor groove surface of the TATA box unwound by TBP. Cell 94: 573–83.

Love, J. J., Li, X., Case, D. A., Giese, K., Grosschedl, R., and Wright, P. E. (1995). Structural basis for DNA bending by the architectural transcription factor LEF-1. Nature 376: 791–5.

Luisi, B. F., Xu, W. X., Otwinowski, Z., Freedman, L. P., Yamamoto, K. R., and Siegler, P. B. (1991). Crystallographic analysis of the interaction of the glucocorticoid receptor with DNA. Nature 352: 497–505.

Ma, J., Sigler, P. B., Xu, Z., and Karplus, M. (2000). A dynamic model for the allosteric mechanism of GroEL. J. Mol. Biol. 302(2): 303–13.

Manna, A. C., Pai, K. S., Bussiere, D. E., Davies, C., White, S. W., and Bastia, D. (1996). Helicase-contrahelicase interaction and the mechanism of termination of DNA repli-cation. Cell 87: 881–91.

Marcotrigiano, J., Gingras, A. C., Sonenberg, N., and Burley, S. K. (1997). Cocrystal structure of the messenger RNA 5′ cap-binding protein (eIF4E) bound to 7-methyl-GDP. Cell 89: 951–61.

Marcotrigiano, J., Lomakin, I. B., Sonenberg, N., Pestova, T. V., Hellen, C. U. T., and Burley, S. K. (2001). A conserved HEAT domain within eIF4G directs assembly of the translation initiation machinery. Mol. Cell 7: 193–203.

Marmorstein, R., and Harrison, S. C. (1994). Crystal structure of a PRP1-DNA complex:

DNA recognition by proteins containing a Zn2Cys6 binuclear cluster. Genes Dev. 8: 2504–12.

Marmorstein, R., Carey, M., Ptashne, M., and Harrison, S. C. (1992). DNA recognition by GAL4: Structure of a protein-DNA complex. Nature 356: 408–14.

Martinez-Yamout, M., Legge, G. B., Zhang, O., Wright, P. E., and Dyson, H. J. (2000). Solution structure of the cysteine-rich domain of the Escherichia coli chaperone protein DnaJ. J. Mol. Biol. 300(4): 805–18.

McCutcheon, J. P., Agrawal, R. K., Phillips, S. M., Grassucci, R. A., Gerchman, S. E., Clemons, W. M., Ramakrishnan, V., and Frank, J. (1999). Location of translation initiation factor IF3 on the small ribosomal subunit. Proc. Natl. Acad. Sci. USA 96: 4301–6.

McKnight, S. L. (1991). Molecular zippers in gene regulation. Scientific American April: 54–64.

Milkereit, P., Gadal, O., Podtelejnikov, A., Trumlet, S., Gas, N., Petfalski, E., Tollervey, D., Mann, M., Hurt, E., and Tschochner, H. (2001). Maturation and internuclear transport of pre-ribosomes requires Noc proteins. Cell 105: 499–509.

Mooney, R. A., and Landick, R. (1999). RNA polymerase unveiled. Cell 96: 687–90.

Morals Cabral, J. H., Jackson, A. P., Smith, C. V., Shikotra, N., Maxwell, A., and Liddington, R. C. (1997). Crystal structure of the breakage-reunion domain of DNA gyrase. Nature 388: 903–6.

Morshauser, R. C., Hu, W., Wang, H., Pang, Y., Flynn, G. C., and Zuiderweg, E. R. P. (1999). High-resolution solution structure of the 18 kDa substrate-binding domain of the mammalian chaperone protein Hsc70. J. Mol. Biol. 289: 1387–403.

Mueller, F., Sommer, I., Baranov, P., Matadeen, R., Stoldt, M., Wohnert, J., Gorlach, M., van Heel, M., and Brimacombe, R. (2000). The 3D arrangement of the 23 S and 5 S rRNA in the Escherichia coli 50 S ribosomal subunit based on a cryo-electron microscopic reconstruction at 7.5 Å resolution. J. Mol. Biol. 248: 35–59.

Muller, C. W., and Hermann, B. G. (1997). Crystallographic structure of the T-domain-DNA complex of the Brachyury transcription factor. Nature 389: 884–8.

Murante, R. S., Henricksen, L. A., and Bambara, R. A. (1998). Junction ribonuclease: An activity in Okazaki fragment processing. Proc. Natl. Acad. Sci. USA 95: 2244–9.

Murray, J. B., Terwey, D. P., Maloney, L., Karpiesky, A., Usman, N., Beigelman, L., and Scott, W. G. (1998). The structural basis of hammerhead ribozyme self-cleavage. Cell 92: 665–73.

Newton, C. S. (1997). Putting it all together: Building a prereplicative complex. Cell 91: 717–20.

Nissen, P., Hansen, J., Ban, N., Moore, P. B., and Steitz, T. A. (2000). The structural basis of ribosome activity in peptide bond synthesis. Science 289: 920–30.

Nolte, R. T., Collins, R. M., Harrison, S. C., and Brown, R. S. (1998). Differing roles for zinc fingers in DNA recognition: Structure of a six finger transcription factor IIIA complex. Proc. Natl. Acad. Sci. USA 95: 2938–43.

Nudler, E., Kashlev, M., Nikiforov, V., and Goldfarb, A. (1995). Coupling between transcription termination and RNA polymerase inchworming. Cell 81: 351–7.

Otwinowski, Z., Schevitz, R. W., Zhang, R-G., Lawson, C. L., Joachimiak, A. J., Marmorstein, R., Luisi, B. F., and Sigler, P. B. (1988). Crystal structure of Trp repressor operator complex at atomic resolution. Nature 335: 321–9.

Pabo, C. O., Aggrawal, A. K., Jordan, S. R., Beamer, L. J., Obeysekare, U. R., and Harrison, S. C. (1990). Conserved residues make similar contacts in two repressor-operator complexes. Science 247: 1210–13.

Pavletich, N. P., and Pabo, C. O. (1993). Crystal structure of a five-finger GLI-DNA complex: New perspectives on zinc fingers. Science 261: 1701–7.

Pazin, M. J., and Kadonaga, J. T. (1997). What's up and down with histone deacetylation and transcription? Cell 89: 325–8.

Pennisi, E. (1997). Opening the way to gene activity. Science 275: 155–7.

Pestova, T. V., Borukhov, S. I., and Hellen, C. U. T. (1998). Eukaryotic ribosomes require initiation factors 1 and 1A to locate initiation codons. Nature 394: 854–9.

Peter, B. J., Ullsperger, C., Hiasa, H., Marians, K. J., and Cozzarelli, N. R. (1998). The structure of supercoiled intermediates in DNA replication. Cell 94: 819–27.

Piper, D. E., Batchelor, A. H., Chang, C. P., Clearly, M. L., and Wolberger, C. (1999). Structure of a HoxB1-Pbx1 heterodimer bound to DNA: Role of the hexapeptide and a fourth homeodomain helix in complex formation. Cell 96: 587–97.

Podobnik, M., McInerney, P., O'Donnell, M., and Kuriyan, J. (2000). A TOPRIM domain in the crystal structure of the catalytic core of Escherichia coli primase confirms a structural link to DNA topoisomerases. J. Mol. Biol. 300(2): 353–62.

Poglitsch, C. L., Meredith, G. D., Gnatt, A. L., Jensen, G. J., Chang, W., Fu, J., and Kornberg, R. D. (1999). Electron crystal structure of an RNA polymerase transcription elongation complex. Cell 98: 791–8.

Polacek, N., Gaynor, M., Yassin, A., and Mankin, A. S. (2001). Ribosomal peptidyl treansferase can withstand mutations at the putative catalytic nucleotide. Nature 411: 498–501.

Polyakov, A., Severinova, E., and Darst, S. A. (1995). Three-dimensional structure of E. coli core RNA polymerase: Promoter binding and elongation conformations of the enzyme. Cell 83: 365–73.

Porse, B. T., and Garret, R. A. (1999). Ribosomal mechanics, antibiotics, and GTP hydrolysis. Cell 97: 423–6.

Powell, L. M., Wallis, S. C., Pease, R. J., Edwards, Y. H., Knott, T. J., and Scott, J. (1987). A novel form of tissue-specific RNA processing produce apolipoprotein-B48 in intestine. Cell 50: 831–40.

Proudfoot, N. (1996). Ending the message is not so simple. Cell 87: 779–81.

Raghunathan, S., Kozlov, A. G., Lohman, T. M., and Waksman, G. (2000). Structure of the DNA binding domain of E. coli SSB bound to ssDNA. Nature Struct. Biol. 7(8): 648–52.

Ramakrishnan, V., and Moore, P. B. (2001). Atomic structures at last: the ribosome in 2000. Curr. Opin. Struct. Biol. 144–154.

Redinbo, M. R., Stewart, L., Kuhn, P., Champoux, J. J., and Hol, W. G. J. (1998). Crystal structures of human topoisomerase I in covalent and noncovalent complexes with DNA. Science 279: 1504–13.

Rice, P. A., Yang, S-W., Mizuuchi, K., and Nash, H. A. (1996). Crystal structure of an IHF-DNA complex: A protein-induced DNA U-turn. Cell 87: 1295–306.

Robinson, H., Gao, Y., McCrary, B. S., Edmondson, S. P., Shriver, J. W., and Wang, A. H. J. (1998). The hyperthermophile chromosomal protein Sac7d sharply kinks DNA. Nature 392: 202–5.

Rodgers, D. W., and Harrison, S. C. (1993). The complex between phage 434 repressor DNA-binding domain and operator site O 3: Structural differences between consensus and non-consensus half-sites. Structure 1(4): 227–40.

Roll-Mecak, A., Cao, C., Dever, T. E., and Burley, S. K. (2000). X-ray structures of the universal translation initiation factor IF2/eIF5B: Conformational changes on GDP and GDP binding. Cell 103: 781–92.

Rould, M. A., Perona, J. J., Soll, D., and Steitz, T. A. (1989). Structure of E. coli glutaminyl-tRNA synthetase complexed with tRNA and ATP at 2.8 Å resolution. Science 246: 1135–42.

Sachs, A. B. (2000). Cell cycle-dependent translation intitiation: IRES elements prevail. Cell 101: 243–5.

Sawaya, M. R., Guo, S., Tabor, S., Richardson, C. C., and Ellenberger, T. (1999). Crystal structure of the helicase domain from the replicative helicase-primase of bacterio-phage T7. Cell 99: 167–77.

Sawaya, M. R., Prasad, R., Wilson, S. H., Kraut, J., and Pelletier, H. (1997). Crystal structures of human DNA polymerase β complexed with gapped and nicked DNA: Evidence for an induced fit mechanism. Biochemistry 36: 11205–15.

Schimmel, P., and Ribas de Pouplana, L. (1999). Genetic code origins: Experiments confirm phylogenic predictions and may explain a puzzle. Proc. Natl. Acad. Sci. USA 96: 327–8.

Schultz, S. C., Shields, G. C., and Steitz, T. A. (1991). Crystal structure of a CAP-DNA complex: The DNA is bent by 90 degrees. Science 253: 1001–7.

Schumacher, M. A., Choi, K. Y., Zaklin, H., and Brennan, R. G. (1994). Crystal structure of the LacI family member, PurR, bound to DNA: Minor groove binding by alpha helices. Science 266: 763–70.

Schwabe, J. W. R., and Rhodes, D. (1991). Beyond zinc fingers: Steroid hormone re-ceptors have a novel structural motif for DNA recognition. Trends Biochem. Sci. 16: 291–6.

Schwabe, J. W. R., Neuhaus, D., and Rhodes, D. (1990). Solution structure of the DNA-binding domain of the oestrogen receptor. Nature 348: 458–61.

Selmer, M., Al-Karadaghi, S., Hirokawa, G., Kaji, A., and Liljas, A. (1999). Crystal structure of Thermotoga maritima ribosome recycling factor: A tRNA mimic. Science 286: 2349–52.

Shamoo, Y., and Steitz, T. A. (1999). Building a replisome from interacting pieces: Sliding clamp complexed to a peptide from DNA polymerase and a polymerase editing complex. Cell 99: 155–65.

Sharkey, M., Graba, Y., and Scott, M. P. (1997). Hox genes in evolution: Protein surfaces and paralog groups. Trends Genet. 13: 145–51.

Sharp, P. A., and Burge, C. B. (1997). Classification of introns: U2-type or U12-type. Cell 91: 875–9.

Shimon, L. J. W., and Harrison, S. C. (1993). The phage 434 OR2/R1-69 complex at 2.5 angstroms resolution. J. Mol. Biol. 232: 826–38.

Shimotakahara, S., Gorin, A., Kolbanovskiy, A., Kettani, A., Hingerty, B. E., Amin, S., Broyde, S., Geacintov, N., and Patel, D. J. (2000). Accomodation of S-cis-tamoxifen-N-guanine adduct within a bent and widened DNA minor groove. J. Mol. Biol. 302: 377–93.

Shippen-Lentz, D., and Blackburn, E. H. (1990). Functional evidence for an RNA tem-plate in telomerase. Science 247: 546–52.

Shyu, A., and Wilkinson, M. F. (2000). The double lives of shuttling mRNA binding proteins. Cell 102: 135–8.

Siegert, R., Leroux, M. R., Scheufler, C., Harti, F. U., and Moarefi, I. (2000). Struc-ture of the molecular chaperone prefoldin: Unique interaction of multiple coiled coil tentacles with unfolded proteins. Cell 103: 621–32.

Singer, R. H., and Green, M. R. (1997). Compartmentalization of eukaryotic gene ex-pression: Causes and effects. Cell 91: 291–4.

Smale, S. T., and Baltimore, D. (1989). The "initiator" as a transcription control element. Cell 57: 103–13.

Sollner-Webb, B. (1991). RNA editing. Curr. Opin. Cell Biol. 3: 1056–61.

Song, H., Mugnier, P., Das, A. K., Webb, H. M., Evans, D. R., Tuite, M. F., Hemmings, B. A., and Barford, D. (2000). The crystal structure of human eukaryotic release factor eRF1-mechanism of stop codon recognition and peptidyl-tRNA hydrolysis. Cell 100: 311–21.

Spronk, C. A. E. M., Bonvin, A. M. J. J., Radha, P. K., Melacini, G., Boelens, R., and Kaptein, R. (1999). The solution structure of lac repressor headpiece 62 complexed to symmetrical lac operator. Structure 7: 1483–92.

Staley, J. P., and Guthrie, C. (1998). Mechanical devices of the spliceosome: Motors, clocks, springs, and things. Cell 92: 315–26.

Stark, H., Orlova, E. V., Rinke-Appel, J., Junke, N., Mueller, F., Rodnina, M., Wintermeyer, W., Brimacombe, R, and van Heel, M. (1997). Arrangement of tRNAs in pre- and posttranslocational ribosomes revealed by electron cryomicroscopy. Cell 88: 19–28.

Stark, H., Rodnina, M. V., Wieden, H. J., van Heel, M., and Wintermeyer, W. (2000). Large-scale movement of elongation factor G and extensive conformational change of the ribosome during translocation. Cell 100: 301–9.

Steitz, T. A. (1992). A general structural mechanism of coupling NTP hydrolysis to other processes. Proceedings of the Robert A. Welch Foundation, Houston, TX, pp. 173–86.

Stewart, L., Redinbo, M. R., Qiu, X., Hol, W. G. J., and Champoux, J. J. (1998). A model for the mechanism of human topoisomerase I. Science 279: 1534–41.

Strobel, S. A., and Cech, T. R. (1995). Minor groove recognition of the conserved G-U pair at the Tetrahymena ribozyme reaction site. Science 267: 675–9.

Su, S., Gao, Y-G., Robinson, H., Liaw, Y-C., Edmondson, S. P., Shriver, J. W., and Wang, A. H-J. (2000). Crystal structure of the chromosomal proteins Sso7d/Sac7d bound to DNA containing T-G mismatched base pairs. J. Mol. Biol. 303: 395–403.

Subramanya, H. S., Doherty, A. J., Ashford, S. R., and Wigley, D. B. (1996). Crystal structure of an ATP-dependent DNA ligase from bacteriophage T7. Cell 85: 607–15.

Tan, S., and Richmond, T. J. (1998). Crystal structure of the yeast MATα2/MCM1/DNA ternary complex. Nature 391: 660–6.

Tan, S., Hunziker, Y., Sargent, D. F., and Richmond, T. J. (1996). Crystal structure of a yeast TFIIA/TBP/DNA complex. Nature 381: 127–34.

Tsai, F. T. F., and Singer, P. B. (2000). Structural basis of preinitiation complex assembly of human Pol II promoters. EMBO J. 19(1): 25–36.

Tuschi, T., Gohlke, C., Jovin, T. M., Westhof, E., and Eckstein, F. (1994). A three-dimensional model for the hammerhead ribozyme based on fluorescence measurements. Science 266: 785–9.

Velankar, S. S., Soultanas, P., Dillingham, M. S., Subramanya, H. S., and Wigley, D. B. (1999). Crystal structures of complexes of PcrA DNA helicase with a DNA substrate indicate an inchworm mechanism. Cell 97: 75–84.

Wang, D., Meier, T. I., Chan, C. L., Feng, G., Lee, D. N., and Landick, R. (1995). Discontinuous movements of DNA and RNA in RNA polymerase accompany formation of a paused transcription complex. Cell 81: 341–50.

Wang, J., Sattar, A. K. M. A., Wang, C. C., Karam, J. D., Konigsberg, W. H., and Steitz, T. A. (1997). Crystal structure of a pol α family replication DNA polymerase from bacteriophage RB69. Cell 89: 1087–99.

Wang, Y., and Patel, D. J. (1993). Solution structure of the human telomeric repeat d[AG (T AG)] G-tetraplex. Structure 1(4): 263–82.

Wei, X., Samarabandu, J., Devdhar, R. S., Siegel, A. J., Acharya, R., and Berezney, R. (1998). Segregation of transcription and replication sites into higher order domains. Science 281: 1502–4.

Weichenrieder, O., Wild, K., Strub, K., and Cusak, S. (2000). Structure and assembly of the Alu domain of the mammalian signal recognition particle. Nature 408: 167–73.

Wells, S. E., Hillner, P. E., Vale, R. D., and Sachs, A. B. (1998). Circularization of mRNA by eukaryotic translation initiation factors. Mol. Cell 2: 135–40.

Wickner, S., Maurizi, M. R., and Gottesman, S. (1999). Posttranslational quality control: Folding, refolding, and degrading proteins. Science 286: 1888–93.

Wilson, K. S., and Noller, H. F. (1998). Mapping the position of translational elongation factor EF-G in the ribosome by directed hydroxyl radical probing. Cell 92: 131–9.

———. (1998). Molecular movement inside the translational engine. Cell 92: 337–49.

Wimberly, B. T., Brodersen, D. E., Clemmons, W. M., Morgan-Warren, R. J., Carter, A. P., Vonrhein, C., Hartsch, T., and Ramakrishnan, V. (2000). Structure of the 30S ribosomal subunit. Nature 407: 327–39.

Wimberly, B. T., Guymon, R., McCutcheon, J. P., White, S. W., and Ramakrishnan, V. (1999). A detailed view of a ribosomal active site: The structure of the L11-RNA complex. Cell 97: 491–502.

Wintjens, R., Lievin, J., Rooman, M., and Buisine, E. (2000). Contribution of cation-π interactions to the stability of protein-DNA complexes. J. Mol. Biol. 302: 395–410.

Xing, Y., Johnson, C. V., Moen, P. T., Jr., McNeil, J. A., and Lawrence, J. (1995). Nonrandom gene organization: Structural arrangements of specific pre-mRNA transcription and splicing with SC-35 domains. J. Cell Biol. 131: 1635–47.

Xu, H. E., Rould, M. A., Xu, W., Epstein, J. A., Maas, R. L., and Pabo, C. O. (1999). Crystal structure of the human pax-6 paired domain-DNA complex reveals specific roles for the linker region and carboxy-terminal subdomain in DNA binding. Genes Develop. 13: 1263–75.

Yudkovsky, N., Ranish, J. A., and Hahn, S. (2000). A transcription reinitiation intermediate that is stabilized by activator. Nature 408: 225–9.

Zhu, X., Zhao, X., Burkholder, W. F., Gragerov, A., Ogata, C. M., Gottesman, M. E., and Hendrickson, W. A. (1996). Structural analysis of substrate binding by the molecular chaperone DnaK. Science 272: 1606–14.

Index

Printed in the United States
by Baker & Taylor Publisher Services